Organization and Pedagogy of Complexity

First published 2023 in Great Britain and the United States by ISTE Ltd and John Wiley & Sons, Inc.

ISTE Ltd
27-37 St George's Road
London SW19 4EU
UK

www.iste.co.uk

John Wiley & Sons, Inc.
111 River Street
Hoboken, NJ 07030
USA

www.wiley.com

Library of Congress Control Number: 2022951290

British Library Cataloguing-in-Publication Data
A CIP record for this book is available from the British Library
ISBN 978-1-78630-704-0

Contents

Preface

This work illustrates some of the fundamental notions introduced in the book *System Architecture and Complexity: Contribution of Systems of Systems to Systems Thinking* (Printz 2020), but it can be read independently since the case studies and reflections presented here are written in an independent manner, without referring to precise aspects of the book in question.

What is difficult to understand in the engineering of these systems, which have relatively recently been named “complex systems” – indeed a proven pleonasm – is that once “integrated”, they appear to be a *whole* that is more than the sum of its parts; but this whole is retroactive on the constituent elements of these parts, thus forming an adaptive cycle which synchronizes the whole with its elements, even as they continue to maintain their autonomy, hence the expression “inseparable yet distinct.”

A system is defined to fulfill certain functions in any given environment constituting the outside of the system, as opposed to its organic elements constituting the inside. The two are connected, but an integrated system must constantly monitor its environment to ensure the latter remains within limits that permit it to continue its nominal functions, that is, what is expected of it as a whole.

In system maintenance operations – there are often decades of useful “life” before the inevitable retirement from its service – whether corrective, punctual or adaptive/progressive maintenance, which is more general, the elements reappear as they are, but we must never forget that they constitute a whole.

systems of the living world, the immune system, without which we would be unable to survive: Philippe Kourilsky, Emeritus Professor of the College de France, a member of the French Academy of Sciences, and the general director of the Institut Pasteur. To live, he tells us, “we must first survive.” He is the author of numerous books including *Le jeu du hasard et de la complexité* (Kourilsky 2014), which makes connections between bioengineering and the engineering of engineers. “Living and surviving”, a systemic cycle between two distinct but inseparable notions, is how we will conclude.

I would like to take this opportunity to sincerely thank my two colleagues, Philippe Kourilsky and Daniel Krob, for the time they have dedicated to my project during the several discussions generated by this book and the previous one.

January 2023

PART 1

The Organic Component of Systems

Introduction to Part 1

This first part illustrates an essential aspect of systems, which could not exist without said material aspect, the hardware (or "organic" as it is called in French), in the sense of the organ allowing concrete action. The entire difficulty of systems is that this component must be the exact mirror of the abstract model set out by the architecture, as recalled in Chapter 1, and reciprocally. Here, we are faced with what physicists call a relationship of complementarity. There are indeed two distinct realities, each with its own logic, but these two realties are inseparable; they are each the reflection of one another. Their correspondence must be maintained in all circumstances, including in the case of a failure following an unexpected event. Every organic element of the system is subject to the environmental constraints of wherever the element is found. It captures information and acts locally thanks to equipment called sensors and effectors, which allow the system to interact with reality while maintaining the service level negotiated with users. The system monitors its environment and monitors itself to ensure that it continues to work properly. In the "engineered" systems discussed in this book, the information required for the "life" of the system circulates electronically or electromagnetically, whatever the scale of the reality it represents, but in reality, several other sources of energy do the work. We must thus define and manage these energetic correspondences with special equipment generically called transducers. This is the principal subject of this first part, for without transducers, the science of systems would be nothing but an abstract game of logic and mathematics, and while this is certainly interesting – fascinating, even – it would not have any real claim on reality. In this game, we will see that that the human element remains preponderant, for it is all part of the creative genius which characterizes the human race, when faced with the many challenges for survival in a world whose complexity is more and more obvious to us.

1

Elements of Systemics: Systems Architecture

1.1. Introduction

This chapter is a summary that my colleague, Professor Daniel Krob, President of the French Center for Excellence in Architectural, Management and Economy of Systems (*Centre d'excellence sur l'architecture, le management et l'économie des systems*, CESAMES), allowed me to reproduce from his lectures and various publications, with some minor adjustments for this present edition. This synthesis contains eight fundamental definitions and a theorem on existence. It is in perfect harmony with my previous book, *System Architecture and Complexity: Contribution of Systems of Systems to Systems Thinking*, from the same collection.

The systems that humans design and make are at the heart of the modern world: it is indeed enough to think of social and political systems, economic and industrial systems, monetary and financial systems, energy production and distribution systems, transport systems, communication systems and more generally, the many organizational and technological systems that surround us, to realize that "engineered" systems are simply everywhere[1]!

More and more, all these systems are characterized by both a *great intrinsic complexity* and ever-increasing interdependencies. Mastering these systems has

For a color version of all the figures in this chapter, see www.iste.co.uk/printz/pedagogy.zip.

1 Living systems in particular will not be considered in this chapter (see the Conclusion, however), which is restricted to engineered systems, in other words, artificial systems (as opposed to natural systems) designed by man, which already covers a vast spectrum, given that technical systems as well as organizational systems can fall into this conceptual category.

DEFINITION 1.1.– *Formal system: a formal system S is characterized on the one hand by the overall data on inputs X, outputs Y and internal states Q, and on the other hand, by the two following kinds of behavior that tie these systemic variables to each other over time:*

– functional behavior, which produces an output $y(t) \in Y$ *at any time t depending on the input* $x(t) \in X$ *and the ordinary internal state* $q(t) \in Q$ *of the system;*

– internal behavior, which causes the internal state $q(t) \in Q$ *of the system to evolve over the course of time t as a result of an input* $x(t) \in X$ *into the system.*

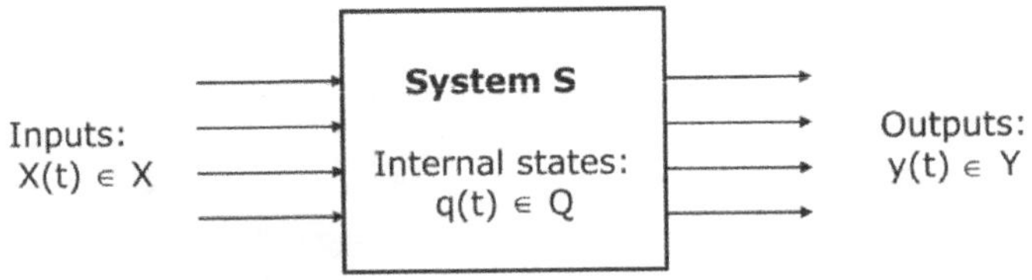

Figure 1.2. *Classic understanding of a formal system*

This definition is intentionally extremely flimsy, for it simply tells us, in essence, that a system is the data of input/output behavior and internal behavior[6]. This is not strange because the purpose is to capture the common points among *all* real systems, which are necessarily very limited given the diversity of objects which we must take into account. On the other hand, we now have a unified framework within which we will be able to reason in a consistent manner about every type of system, and thus theorize integration, which was the first of our objectives (see also, D. Krob, *Elements of Complex Systems Architecture*)[7].

DEFINITION 1.2.– *Real system: an object from the real world is called a real system from the moment its structure and behavior can be described by a formal system (in the sense of Definition 1.1), which we then call a model of the real system.*

The near totality of human inventions – whether they be of an electro-mechanic, computing, or organizational nature – are thus analyzable as systems: the only change that occurs is in the nature of the laws – in this case physical, logical or

6 Modeled by a law of evolution of its internal variables, which we call "states" here.

7 In Appriou, A. (ed.) (2009). *Gestion de la complexité et de l'information dans les grands systèmes critiques*. CNRS Editions, 179–207. See also *Model-based Systems Architecting*, ISTE-Wiley, 2022.

sociological – which we use to define the functional and internal behaviors that make them into systems. The previous definitions also underscore that it is neither the nature, nor the size, nor the hierarchical position that make up a system, since almost every object can be seen as a system[8]. Considering an object as a system rests first and foremost on the choice in modeling: it is indeed the same as considering a particular object in the abstract to restrict the analysis to the systemic perspective, that is, within the framework given by Definition 1.1.

Let us point out that from here on, we will only provide definitions for formal systems, bearing in mind that they transfer directly onto real systems, given the modeling perspective that we have adopted here.

1.2.3. *Integration of systems*

We are now equipped to introduce the notion of system integration, which is the fundamental mechanism that allows, in practice, for the construction of a new system using other, smaller systems (materials, software, humans) by organizing them in such a way that it can execute – within a given environment – its assigned tasks.

DEFINITION 1.3.– *Integration: let S_1, ..., S_N be a set of N (formal) systems. We can thus say that a (formal) system S is the result of the integration of these systems if there exists, on the one hand, a (formal) system C obtained through the composition of systems S_1, ..., S_N, and on the other hand, dual abstraction and concretization mechanisms that let us express:*

– *system S as an abstraction of system C;*

– *system C as a concretization of system S*[9].

This somewhat technical definition has the fundamental objective of positing integration as a mechanism that is distinct from the simple composition of models, in order to try to capture the emergence in our approach, structurally.

8 A notable common error is to assume that a "sub-system" of a system (see section 1.2.3 for more details) is not a system, which is naturally not the case (and which lets us apply the principles of systemic analysis to this level too).

9 We will not specify the notions of abstraction and concretization here, which should be taken in the sense of the theory of abstract interpretation (see, for example: Cousot, P., Cousot, R. (1996). Abstract interpretation. *Symposium on Models of Programming Languages and Computation, ACM Computing Surveys*, 28(2), 324–328, for more details on this subject).

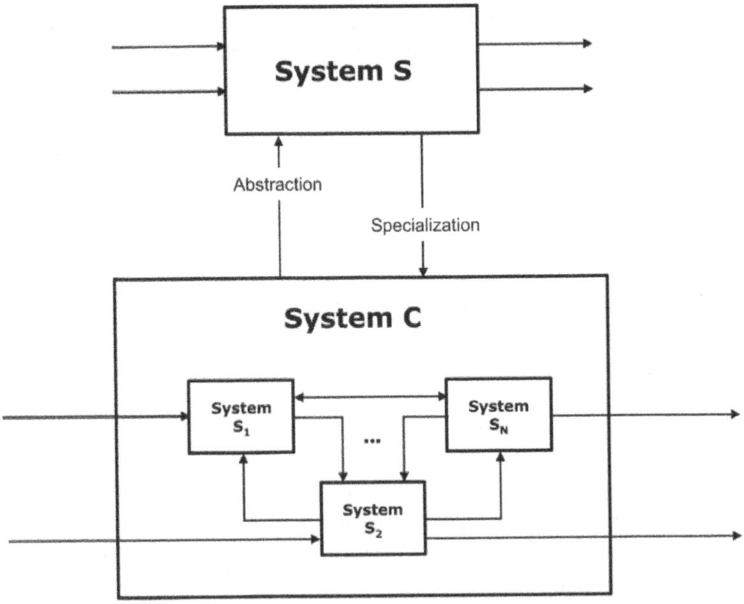

Figure 1.3. *Formal integration of (formal) systems*

Our definition indeed clearly presents the fact that the description of a high-level system obtained through the interconnection of other systems requires us to have:

– models of each of the systems that constitute the high-level system;

– a new, high-level model that is specific to this system.

In other words, simply knowing the models of the constitutive parts of a system is never enough to model this system! This modeling postulate can be seen as the direct translation of the universal phenomenon that is emergence. It can moreover be verified within all the elementary systems of everyday life.

Let us consider, for example, a wall made only of bricks (without mortar to hold them together), to simplify things. A simplified systemic model of a brick can thus be characterized by:

– functional behavior consisting of the absorption of rays of light;

– internal behavior specified by the constant states of "length, width, height."

In making a model of bricks, it is thus difficult to obtain anything other than the functional behavior of absorbing rays of light, which is clearly not the usual behavior of a wall since we want to be able to reflect the holes that exist in walls (windows), which themselves actually let the light through! We will also note that it is difficult – and surely in vain – to try to express the form of a wall (which is one of these typical internal states) in terms of the lengths, widths and heights of the bricks that make it up. All of this is therefore an argument for the fact that we must clearly create a dedicated, more abstract, systemic model in order to model the wall.

These observations perhaps appear to be merely common sense, if not naïve, but we unfortunately see that their consequences are more often than not misunderstood, which is the direct source of many of the quality problems observed in the great engineered systems of modern times. In fact, most people in the industry continue to think that mastering the components of an integrated system is sufficient for mastering the system as a whole. But the very nature of the mechanism of integration requires us to go beyond the satisfaction of managing components, when what we should be doing is appreciating it as an integrated system: it is just as necessary to have someone specifically managing the integrative model of the high-level system, that is, a systems architect, who is in fact usually missing from industrial structures, however strange that may appear[10]!

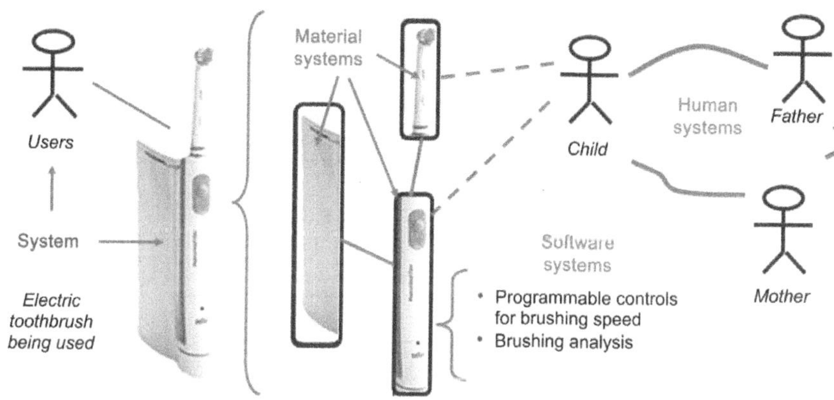

Figure 1.4. *Concrete integration of (real) systems*

Finally, we should point out that the integration mechanism naturally creates hierarchies in systems, while simultaneously giving the systemic analysis a recursive dimension. Engineered systems are indeed made through successive system integrations: any integrated system therefore creates an integration hierarchy, which is also a hierarchy in abstraction given the nature of integration, and is called the

10 The high-level models of integrated systems also often being consequently almost inexistent, in practice.

systemic hierarchy of the initial system, which allows us to speak of the levels, sub-systems, sub-sub-systems, etc. of an integrated system.

1.3. Organization of a systemic model

1.3.1. *Architectural visions of a system*

The systemic approach consists of analyzing real systems as formal systems, which are therefore models that need to be identified through a systemic lens, in which we only think in terms of the interacting systems. The recursive character of systemic analysis thus automatically leads us to introduce the notion of the system environment, which denotes a closed super-system (i.e. without inputs or outputs) containing the system, therefore acting as the natural starting point for the process of recursive system analysis (see the following section).

DEFINITION 1.4.– *Environment of a system: let S be a (formal) system. We can therefore say that a (formal) system E(S) is an environment for S if it is a (formal) closed system*[11] *that results in the integration of S and another (formal) system Ext(S), which we call the exterior of S in the environment E(S).*

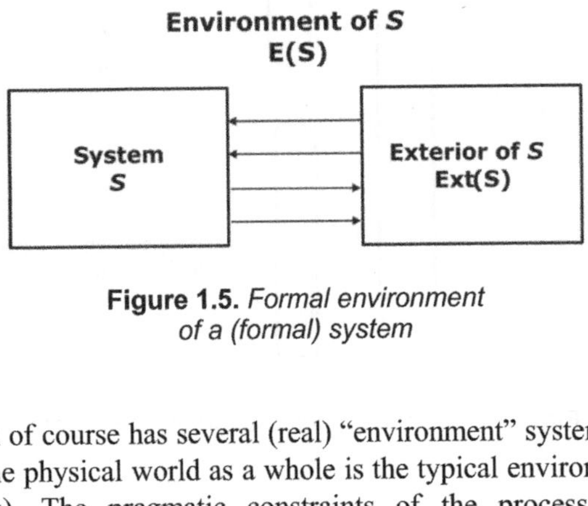

Figure 1.5. *Formal environment of a (formal) system*

A real system of course has several (real) "environment" systems in the sense of Definition 1.4 (the physical world as a whole is the typical environment common to all real systems). The pragmatic constraints of the process of systemically conceiving of a given (real) system S lead us however to introduce the reference environment for S (which is often simply called the environment of S, in short), corresponding to the smallest "useful" environment of S. This is in fact none other than the system resulting from the integration of system S with all the

11 That is, a system whose sets of inputs and outputs are empty, which is the same as saying that there are no inputs or outputs.

(real) systems external to S that have an influence on its conception, consequently ignoring all the other (real) systems external to S when they are considered to have no interdependence (and thus no functional interaction) with the system being envisaged[12].

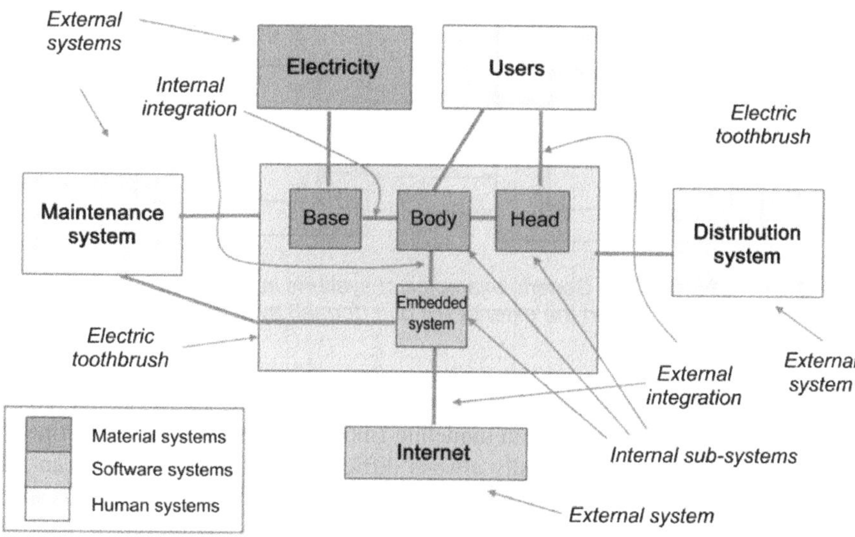

Figure 1.6. *Example of the reference environment of a real system*

We should point out that from this point on, everything that follows will be relative to a chosen environment for a system that will therefore always be considered as fixed a priori, which will let us speak without misnomers about the "environment" of a system. We can thus naturally adopt the following definitions, which model the fact that the environment of a system S automatically defines a boundary between the interior and the exterior of S.

DEFINITION 1.5.– *Interior and exterior: let S be a (formal) system, E(S) its (formal) environment, and Ext(S) its (formal) exterior in this environment, in the sense given in Definition 1.4. We can thus say that a (formal) system I (with respect to E) is in the interior (with respect to the exterior) of S if system S (with respect to system Ext(S)) can be obtained by integrating I (with respect to E) and another (formal)*

12 We can in fact reason by considering that Proxima Centauri has no influence on most earthbound engineered systems.

We can first note that Definition 1.6 is indeed consistent, since the existence of non-organic functional models is directly tied to the postulate of emergence (see section 1.2.3), which, let us recall, confirms that the mere knowledge of the concrete components of a system and their laws of interaction is never enough to describe the behavior of the integrated system resulting from their interconnection. This explains why there are still purely functional models of systems whose role is to clearly express the emergent functions that we would not be able to see at the organic level[13].

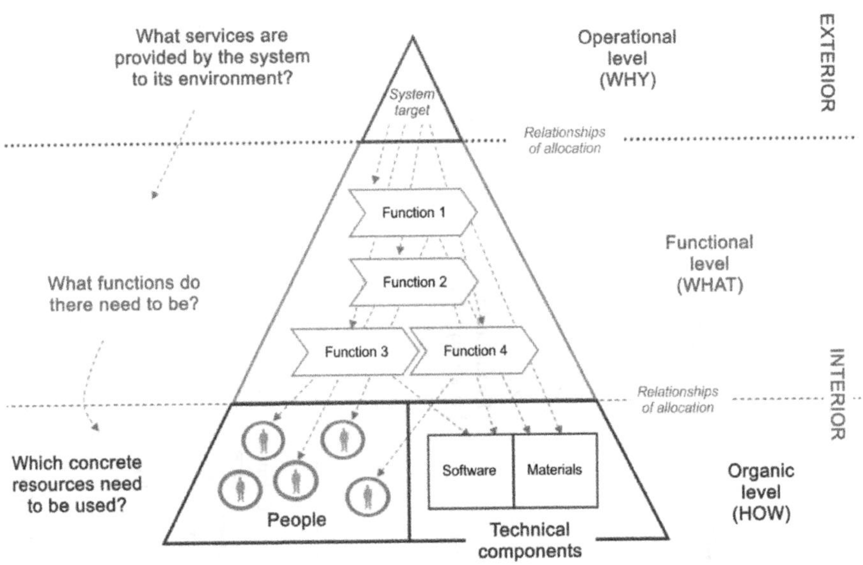

Figure 1.9. *A universal framework for the architectural analysis of a real system*

13 In order to understand this phenomenon well, let us take the example of a car whose (high-level) systemic constitutive components are the undercarriage, the motor, the passenger compartment, the contact with the ground and the built-in electronics. The interaction of these different components is typically sufficient for the performance of functions such as the detection of obstacles, which requires the cooperation of a radar (placed in the undercarriage), an accelerometer (part of the built-in electronics), an illuminated indicator (placed in the passenger compartment) and even the contact with the ground or the motor if we want to act upon the brakes and/or reduce the engine torque when an obstacle is too close to a car. Such a function – called transversal – would clearly be difficult to capture in a purely organic model of the car, where we see exchanged flows between the different components of the vehicle without being able to explain their overall logical structure. Only a functional model of the whole car can therefore capture the semantics of such a transversal function.

We now also understand why it is necessary to have three types of models to model a real system in practice: the operational vision captures the external point of view, while the functional and organic perspectives capture the internal point of view by respectively modeling the emergent behaviors on the one hand, and, on the other, the concrete constitution of the system being considered.

With this now elucidated, it is useful to underscore that the architectural visions play a key role in systems architecture: the first job of a systems architect will as matter of fact consist of classifying the modeling elements at hand, which most often, in practice, include various conceptual levels, according to the classification from Definition 1.6, in order to obtain homogeneous models, each capturing a very limited perspective.

1.3.2. *Properties of a system*

The first way of systemically analyzing a system consists of describing the properties that must be satisfied, which gives rise to the logical models describing the systems that these requirements constitute (which are typically used, in practice, as specification tools for writing requirement specifications in engineering).

DEFINITION 1.7.– *Requirement: let S be a (formal) system. Any logical property (temporal or not) that can be expressed in the language of S, that is, by way of its inputs, outputs and states, can thus be called a requirement of system S.*

To say that an airplane is safe, for example, is in essence the same as imposing a safety requirement upon the "airplane" system, affirming that there is no moment in time in which this system is in a state of "catastrophe" (which must naturally be defined precisely and integrated into the model of the airplane in advance in order to be able to express the previous requirement of safety).

We can now identify how this generic notion of a requirement overlaps with all the architectural perspectives defined in the previous section, which brings us to Definition 1.8.

DEFINITION 1.8.– *Need, functional requirement and organic requirement: let S be a real system. We can thus say that a logical property is:*

– *a need if it is a requirement of an operational model of S;*

– *a functional requirement if it is a requirement of a functional model of S;*

– *an organic requirement if it is a requirement of an organic model of S.*

These three concepts are fundamental since the entire architectural reasoning consists of starting from needs, that is, the properties expected by the exterior of the system (what the "clients" of the system want), in order to translate them into functional requirements (what the system must do), and then into organic requirements (how the system must concretely be made).

In practice, organizing the properties that appear empirically throughout the engineering process according to the categories in Definition 1.8 presents major difficulty. It is indeed very easy to identify properties that merrily blend the three architectural perspectives: from the point of view of the constructor of military vehicles, the property "soldiers must move around in a transport vehicle for troops that is painted in green" is thus not a need, nor a requirement of the "transport vehicle" system as it combines external elements (soldiers) and internal elements (the color of the vehicle). Architectural reasoning demands us to rewrite this property in the form of an organic requirement: "the vehicle must be the color green," which then immediately raises the question of what the (unspoken) need of the soldiers was. We can easily understand with this example that a functional requirement for the vehicle is missing, namely, "the vehicle must be invisible," which itself is derived directly from a very simple need that the soldiers have: "the soldiers must not die during operations!" By proceeding with this reasoning, we thus make explicit the role of the field of operations in the environment of the target transport system, which in turn lets us realize that the choice of green paint is perhaps well adapted to the wooded terrains of Europe, but surely much less so in the sands of the Gulf, where an ochre color or camouflage would be much more efficient in covering the primitive need of the soldiers, now made clearly explicit.

1.3.3. *Descriptions of a system*

The second way of analyzing a system systemically consists of explicitly describing its behavior in the form of symbolic descriptions using a (formal) ad hoc syntax or, in other words, making its "blueprint."

The architectural perspectives introduced in Definition 1.5 thus let us posit the following definitions.

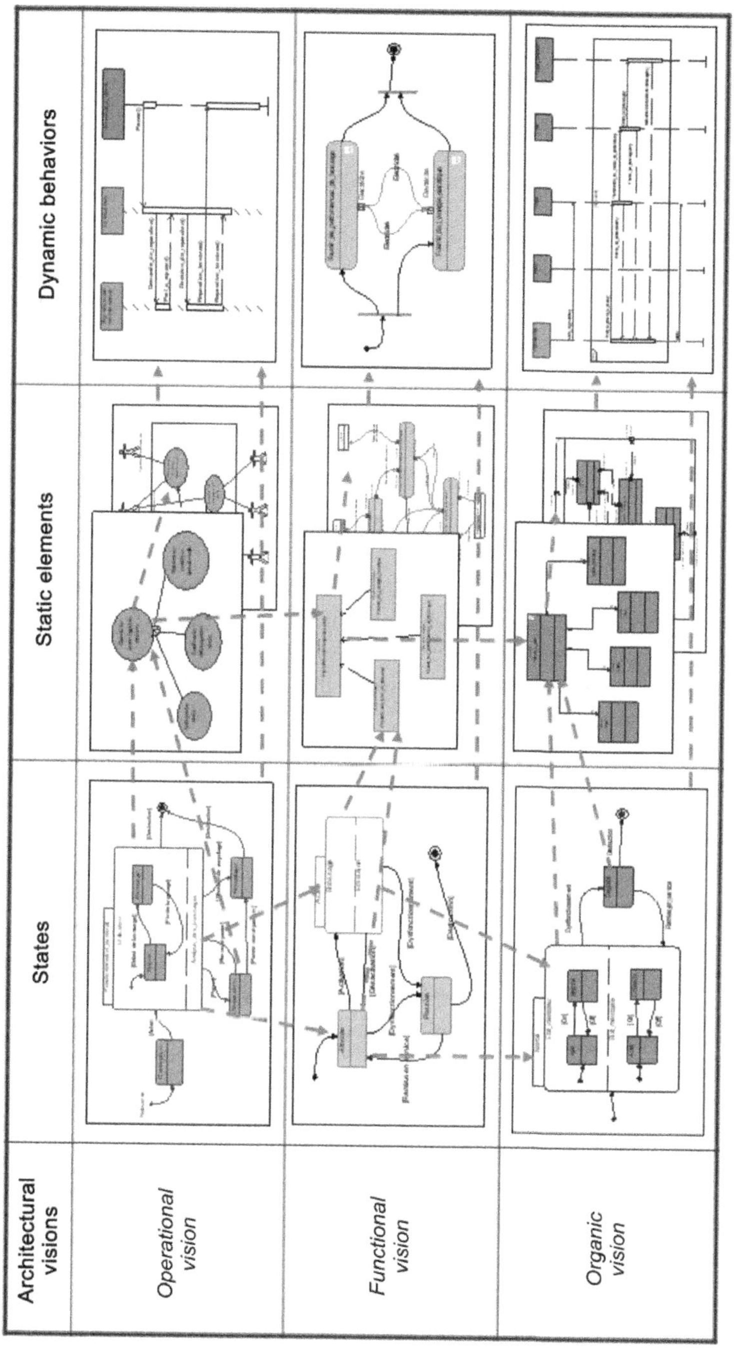

Figure 1.10. *Sample organization of real system descriptions*

DEFINITION 1.9.– *Operational, functional and organic descriptions: let S be a real system. We can thus say that a symbolic representation of this system is:*

– an operational description if it is part of an operational model of S;

– a functional description if it is part of a functional model of S;

– an organic description if it is part of an organic model of S.

We should take note that the very definition of a (formal) system naturally allows us to organize each description of a system according to the following axes, which convey all of the possible components of such a (formal) model of the system:

1) a (static) representation listing states;

2) a (static) representation listing inputs/outputs;

3) a representation of the dynamics of the evolution of states;

4) a representation of the dynamics of the behavior of inputs/outputs.

For practical reasons, we usually group the representations in categories (1) and (3) into just one representation – in the form of a finite-state machine – focused on states, giving both their list and their evolution dynamic, and we break representation (4) into two representations, the first of which simply lists the components necessary for the second in a static manner, focused on modeling the behavior dynamic of the system. Figure 1.10 illustrates an example – by abstracting the representation of inputs/outputs – of this way of organizing the descriptions of a system.

1.3.4. *Reference framework for system analysis*

In summary, we have thus clearly shown that a real system can be analyzed according to the three angles given by its architectural perspectives (operational, functional and organic), its properties and its descriptions (see Figure 1.11).

If we also integrate the fact that a system additionally has a systemic hierarchy (its system levels, sub-systems, sub-sub-systems, etc.), we understand that a system is a fundamentally multidimensional object that is structurally difficult to analyze since we must understand not only each of its architectural components, but also the entirety of their relationships!

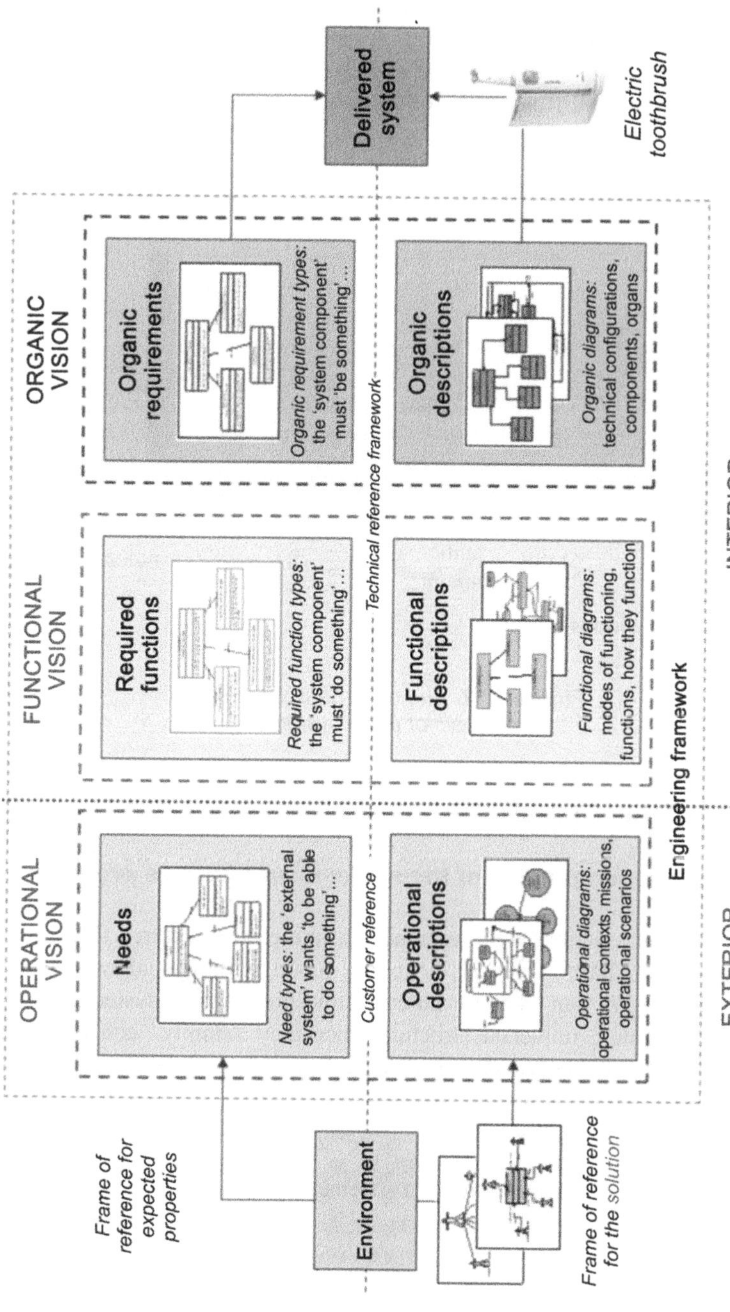

Figure 1.11. *The angles of analysis of a (real) system*

Finally, we should note that all our considerations can be summarized in the architectural analysis cube presented in Figure 1.12, which provides a generic organization of a *systemic model* of a real system and therefore necessarily contains the *reference framework of the systemic analysis* (which consists of analyzing a real system according to the different axes of the cube).

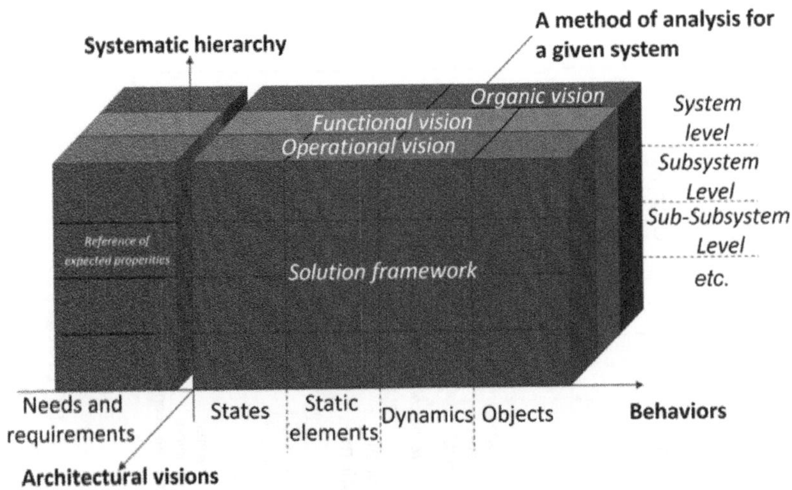

Figure 1.12. *The cube of architectural analysis of (real) systems*

1.4. Architecture of a system

1.4.1. *Systemic perspective of the systems architecture process*

In the previous section, we presented the universal structures that exist in all systems which, in particular, led us to the mode of universal organization for systemic models. We can likewise note that the problems in systemic conception also have a common, universal structure since they "simply" consist of making constructive proofs of theorems that (theoretically) all have the following form[14].

THEOREM 1.1.– *There exists a system S that satisfies a set B of needs.*

Equation 1 – Universal structure of a statement of systemic conception.

14 That is, by concretely building a system that verifies the theorem that needs to be proven.

We thus understand more easily why the existence of a universal identification and construction algorithm for systemic models, which we can call the systems architecture process, is not at all abnormal. It is this latter process that we will now quickly present in section 1.4.2.

Let us first note that the systems architecture process is the same as the functional perspective of the organization in charge of this process, in other words, the functional trace of an architectural conception system that can therefore be systemically analyzed (see Figure 1.13). This "strange cycle" – very characteristic of the reflexivity of systemics which can make use, like mathematics, of its own tools for studying its internal functioning – lets us use the reference framework of architectural analysis, introduced in the previous section, to describe the systems architecture process, as we will now do[15].

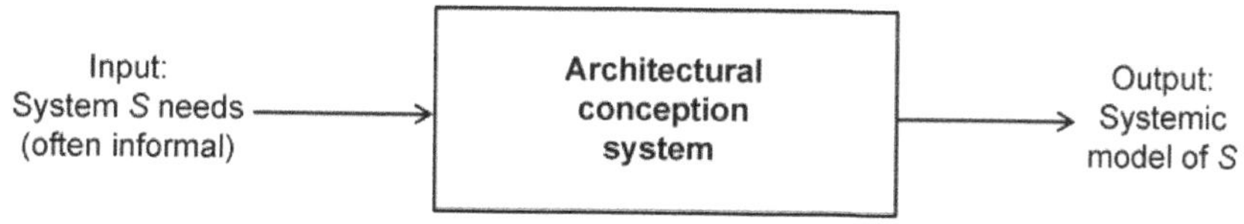

Figure 1.13. *Systemic perspective of the architectural conception system*

1.4.2. *Architectural perspectives of the systems architecture process*

As we underscored above, any method of systemic analysis first consists of identifying the environment of a system. In the case of an architectural conception system, it is none other than the environment of the engineering process itself, of which the systems architecture process is a component: the latter is typically formed of the company in which the process is deployed, its competitors and clients, the users, operators and suppliers it interacts with, and the regulations and technologies it must account for.

Performing an operational analysis of the architectural conception system thus means asking ourselves what the service that the systems architecture process provides its environment with is; recall simply that systems architecture seeks to master the conception and evolution projects of systems, respecting their cost, deadline, quality and performance constraints. Seen from the outside, the

15 See the book Hofstadter, D.R. (1979). *Gödel, Escher, Bach: An Eternal Golden Braid*, Basic Books, which also exists in a French translation.

architectural conception system therefore has the principal objective of helping the actors in a project making or transforming a system to always act in a synchronized manner, the absolute condition for succeeding in any integration, by leaning on universal, exhaustive, coherent and shared models.

At the *functional* level, the activities of the systems architecture process are very simple to describe since they automatically derive from the reference framework of the systemic analysis, introduced in the previous section. These can indeed be separated naturally into the following activities:

– the architecture of a (real) system's needs and requirements, with the objective of formalizing its functional and organic needs and requirements;

– the operational, functional and organic architecture of a (real) system, with the objective of building its operational, functional and organic models;

– verification and validation, which consist of moving up the cycle in order to ensure the real system actually satisfies its requirements and needs.

The key point that should be underscored here is that we can associate a generic process (that would of course be too long to describe here) to each of these architectural activities, which simultaneously allows for the universal method of system conception that we spoke of earlier to emerge, along with a "toolkit" of systems architecture[16]. It is also interesting to give the overall dynamic that exists between these different activities (see Figure 1.14). Finally, let us finish with the organic dimension of the architectural conception system. It is made up of two primary components:

– *humans* including, notably, the key character of the systems architect, who has a double role in managing the integration of the technical systems and human systems in their field of charge: building coherent interfaces within a technical system always in fact requires bringing the interested parties of these interfaces around to a unified vision of their system;

– the *tools* of the systems architecture process, which on the one hand include software tools for the engineering requirements and behavior modeling, based on description languages such as BPMN (Business Process Modeling Notation), UML (Unified Modeling Language) or SysML (System Modeling Language), and on the other hand, the conceptual tools introduced in section 1.3.3, not to mention common

16 See Printz (2019), which exists in an English translation.

sense and heuristic rules, of course, as well as the good practices that come from experience[17].

1.5. In conclusion

We hope that this summary has allowed the reader to better understand the stakes and issues involved in systemics. It naturally does not have the pretension of covering the entirety of this discipline and intentionally omits numerous very important subjects: systems dynamics, systems sizing and optimization, systems evolution trajectories, the conception of homogeneous families of systems, agile and collaborative approaches to systems development, systems governance, etc., which are covered in the previously mentioned book, *Systems Architecture and Complexity*, and the others mentioned throughout the following chapters and in the references. The world of systems is indeed a "continent" that forms a real scientific field for a science of systems[18].

Systemics still however needs to be developed, necessarily through a profound interaction between theory and practice, despite realizations that are, even now, old – and unfortunately almost without follow-up – by the great fathers of the field such as Simon[19] or von Bertalanffy[20]. Let us therefore hope that the systemic challenges that the world is beginning to face, with the financial crisis of 2008, then the much more profound one caused by the emergence of the Covid-19 virus, will finally allow the field to fully emerge[21].

---

17 On BPMN, see White, S.A. and Miers, D. (2008). *BPMN Modeling and Reference Guide, Understanding and Using BPMN.* Future Strategies, as well as the general bibliography. On UML, see Booch, G., Jacobson, I., Rumbaugh, J. (2004). *The Unified Modeling Language Reference Manual*, 2nd edition. Addison-Wesley, as well as the general bibliography. On SysML, see Friedenthal, S., Moore, A.C., Steiner, R. (2012). *A Practical Guide to SysML: the Systems Modeling Language*. Morgan Kaufmann OMG Press.

18 To use the expression that Marcel Paul Schützenberger surely would not have rejected (founding "father" of the French school of theoretical computer science, with Maurice Nivat).

19 See Simon, H. (1962). The architecture of complexity. *Proceedings of the American Philosophica*, 106(6), 467–482, as well as the general bibliography.

20 See von Bertalanffy, K.L. (1968–1976). *General System Theory: Foundations, Development, Applications*, George Braziller, a French translation has been published with Dunod, *Théorie générale des systèmes*, 1973. For a detailed study of the historical aspects, see the thesis Pouvreau, D. (2013). "Une histoire de la 'systémologie générale' de Ludwig Bertalanffy – Généalogie, genèse, actualisation et postérité d'un projet herméneutique", EHESS, March.

21 See De Weck, O., Krob, D., Lefei, L., Lui Pao, C., Rauzy, A., Zhang, X. (2020). Handling the COVID-19 crisis: Towards an agile model-based systems approach. *Systems Engineering*, 23(5), 656–670.

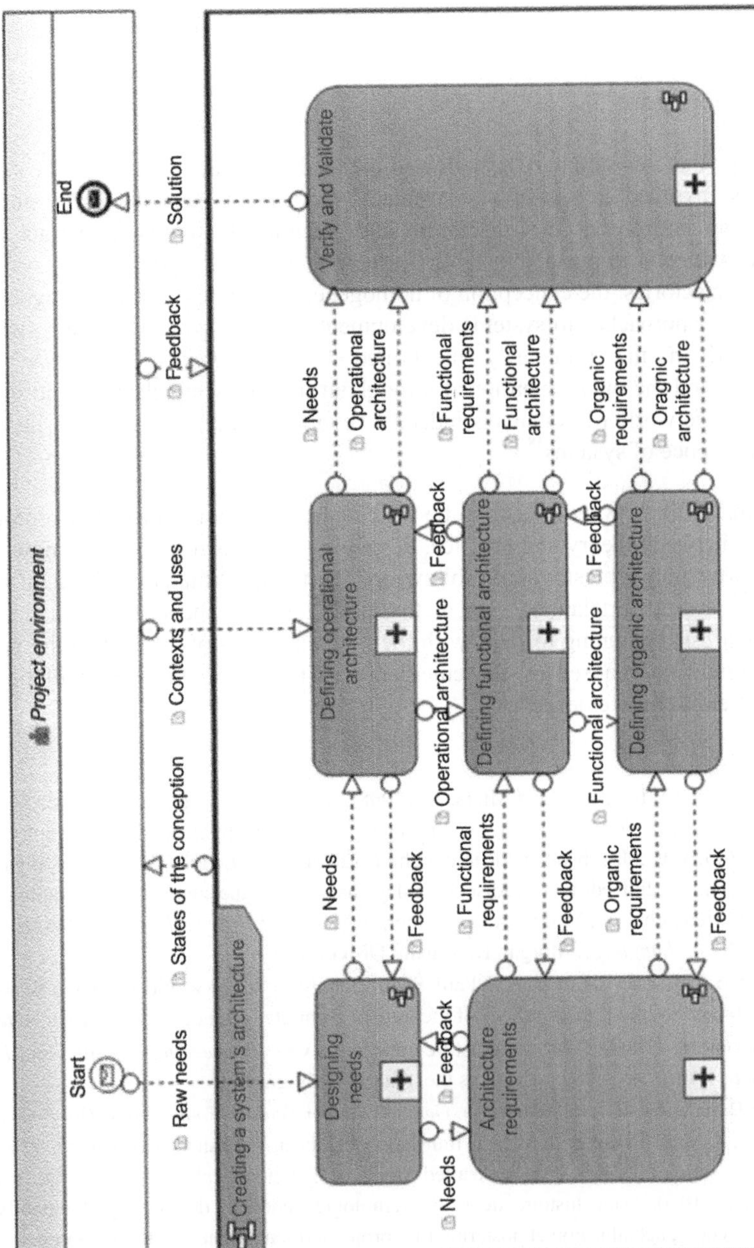

Figure 1.14. *Description of the functional dynamic of the architecture process of systems*

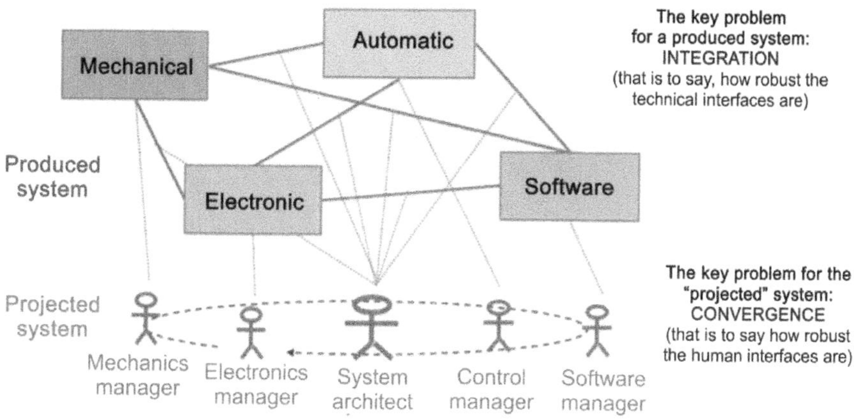

Figure 1.15. *The role of a systems architect*

2

Natural Functions

2.1. The notion of energetic transducers, revisited

In the systems literature in general, and in the aforementioned book, *System Architecture and Complexity: Contribution of Systems of Systems to Systems Thinking*, we often use the terms "transducer" and/or "transduction", which come from physics, and more specifically, from electronics. They are also used in linguistics, often in relation to the more familiar term "translation", as well as in biology for organelles such as ribosomes, which "translate" the genetic information in DNA into amino acids, which in turn make up the proteins that are the basic building blocks of the living.

It is important to clearly grasp the nuance in transduction/translation, as well as the way they are used in systemics and in this book.

A transducer is a physical or physio-chemical system (but we can also speak of a transducer component or element, depending on the level of abstraction in which we find ourselves) that provides a useful physical measure (i.e. a finalized measure with a precise objective) as an exit signal in response to another specified physical measure, such as an entry signal. Anyone who handles a touchscreen device makes use of transducers.

In theory, transduction is a phenomenon of synchronous energetic conversion that conserves the information carried by the entering physical measure. It is a process for which we could establish a conversion table, "Entry × Exit", depending on the physical phenomena implemented: mechanics, optics, chemical, electric, magnetic and thermal. The diagram for the principle of transduction is presented in Figure 2.1.

For a color version of all the figures in this chapter, see www.iste.co.uk/printz/pedagogy.zip.

We also know how to do this, thanks to human engineering, which we control at every level, such as the electric system or in the LHC at CERN, with perhaps the only reservation being that the weight of the instrument's detectors[2] amounts to tens of thousands of tons, and is thus a very far cry from nature's ability, but anyhow…

When we interact among ourselves via human languages, we operate a mechanism of transduction based on shared experiences, which allows us to collectively act in a coherent manner. This chain is a perfect illustration of the dichotomy of transduction/translation, as illustrated in Figure 2.3.

At the center of the mechanism, the human brain encodes/decodes continuous information (ambient noise, coupled with vision or other sensations) according to modalities that are specific to each of the 6,000 languages still used on the planet. What is remarkable in this mechanism is the relative independence of interpretation/action (functional capacity, which seems to come purely from logic; see the work by Noam Chomsky[3], and the *Cours de linguistique générale* by Ferdinand de Saussure) with respect to the hearing and speaking organs themselves. These organs can change or present with local specificities, such as accentuation, without impeding understanding, up to a certain point. Moreover, the brain of a baby has the capacity to learn any language, provided that this capacity is correctly stimulated in the first years of life, and even two languages simultaneously for bilingual couples. The lack of interaction progressively and irreversibly eliminates this capacity over time.

Returning to the reality of systems whose engineering we do master, any "translation" presupposes a transduction exercising one or more energetic phenomena that must be coordinated and controlled so that interpretation/action always remain possible, independently of the context. Rigorous separation of a) the logical–functional plane, which is what allows for coherent interaction, and b) the organic plane, which includes possibly unstable energetic phenomena via ad hoc interfaces, is the essence of what a systems architect and a scientific engineer does. The synchronous/asynchronous dichotomy must also be dealt with to avoid incoherence, given that the human brain is apparently capable of working alongside mechanisms that we currently know nothing about. In the real world, it is strictly impossible to "forget" the underlying organic realm, whose proper energetic operation we rely upon for interpretation and shared meaning. In the new world of the first modern computers, this is what von Neumann would call the *logical design* or the *logical model*.

2 See: http://home.web.cern.ch/fr/about/accelerators/large-hadron-collider.

3 See Berwick, R.C. and Chomsky, N. (2017). *Why Only Us – Language and Evolution*. MIT Press, Cambridge, MA.

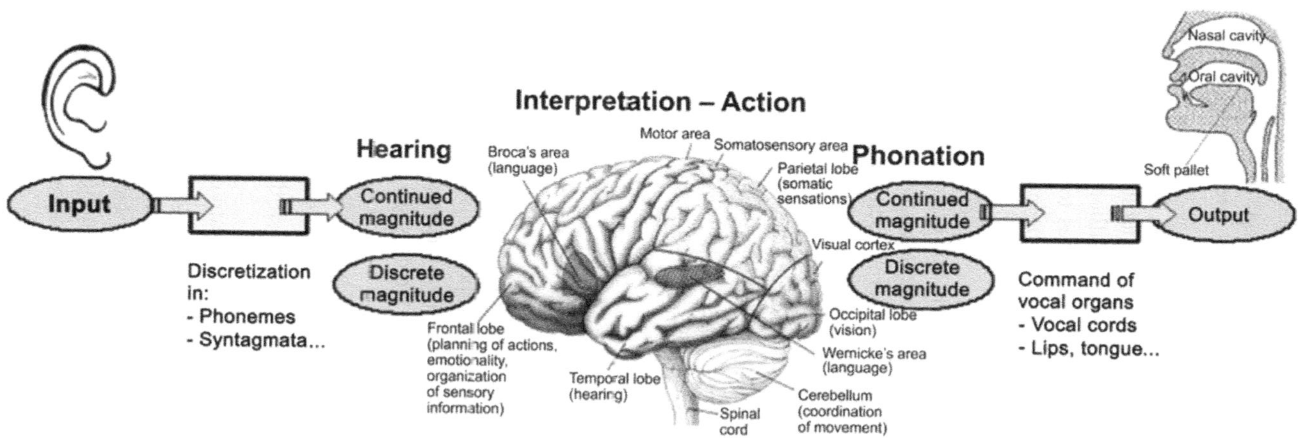

Figure 2.3. *Cerebral/aural/phonation transducer*

This logical model constitutes the fundamental constant of the system; moreover, the elements that constitute the organic "center" of the system rely on its robustness and resilience in order to maintain their properties, and if this is not the case, to stop them promptly and/or replace them without interrupting service, if possible.

The electronic mechanisms of different communication networks that envelop the planet are an excellent example of the successful integration of a very large number of transducers of every kind (millions of kinds of equipment), of every strength, allowing humans and/or machines to interact efficiently according to the tasks set out for them and/or their specific needs, with the objective of ensuring social order worldwide.

Transducers, whatever their nature and organic functions, are the key elements of this mechanism of overall proper systems operation. This is why they must be at the center of a systems architect's concerns. Behind every interface, there is always a transducer, and there is also always a hidden language, even if it does not reveal its name; but it is there, it exists! Therein lies the hidden meaning of the slogan "It from bit", proclaimed by the physicists of the Santa Fe Institute, including, among others, Wheeler, the author of this formula, since information is essentially "language", or, if you like, "structure" or even "organization", which brings us back the patient search for these languages and the abstractions that define them.

2.2. Some fundamental transducers

Historically, the most interesting transducers were, at first, the ones that calculated the common operations of arithmetic, which we will discuss again later, as well as the ones that calculated operations of differential and integral equations, or calculated the derivative of a function represented by a sign as a function of time, or an integration, thanks to components such as capacitors and/or self-inductance coils.

From a general point of view, the transducer was able to produce transformations, in the mathematical sense of the term (see Figure 2.4).

All the extrema of the curve have derivatives of zero, and all the inflection points have second derivatives of zero, which can give us an initial idea of the form of the derived curve that the transducer will "calculate" at the speed it takes its circuits to stabilize once it is powered; this is more efficient in analog than in digital. As for the part that is filled in, it represents a summation; in the case of the figure, we are instead dealing with "energetic" summations that correspond to a Lebesgue integration, more general than the Riemann one. We can see why. Wiener was so interested in all of this.

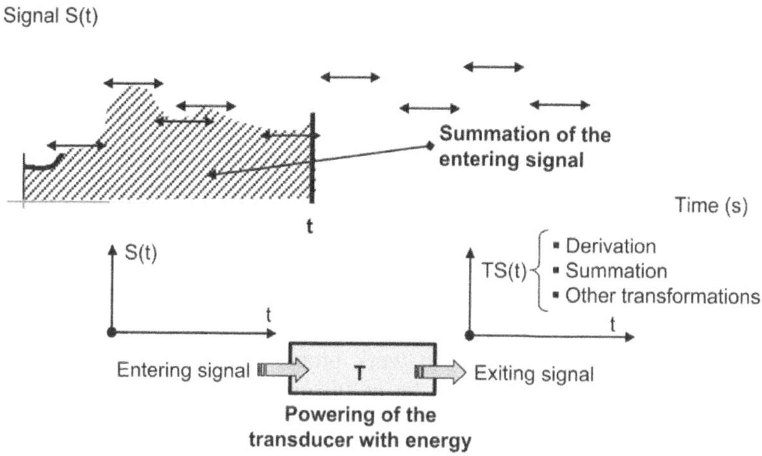

Figure 2.4. *Transformation transducer for an energetic signal*

Without having a lesson on signal processing, we can intuitively understand that these transformations, coupled with periodic functions, have made transducers a fundamental element in the construction of systems such as the ones imagined by. Wiener and his colleagues in electronic and automation engineering at MIT.

But for neurologists such as Walter McCulloch, or the Harvard psychologists who studied the sensations we feel via our senses, there was something better. In psychophysics, there are entirely surprising laws which stipulate that sensation varies as if it were the logarithm of excitement, that is, the quantity of energy captured by the senor. These are the Weber–Fechner laws, or the Stevens laws, named for the eminent Harvard professor (1906–1973) whose *Handbook of Experimental Psychology* (1951) is still used for everything dealing with ergonomics and human–machine interfaces. We know that logarithms and power laws are two ways of expressing the same physical reality. Stanley S. Stevens preferred the formulation using power law, which he denoted by $Sensation \rightarrow \Psi(I) = k \times I^{\alpha}$ (with the letter Ψ for psyche): laws whose form is well known in many areas of engineering and that we also find in numerous natural phenomena; in its logarithmic form, it is written: $Log[\Psi(I)] = Log(k) + \alpha \times Log(I)$, which allows us to say that sensation varies according to the log of excitement. For example, geologists have noticed that the frequency and force of earthquakes in a seismic region such as the San Andreas fault line in California follow a power law[4].

4 For an example of use in the field of project engineering, see Printz (2013).

The truly interesting question is how we understand the structure that is hidden "behind" a power law, since it is almost "too good to be true". For that, let us imagine a surface absorbing energy, for example a surface that is sensitive to heat transfer. The sensation felt depends on two factors:

– the thermal shock due to the difference in temperature (a gradient);

– the size of the receptor surface absorbing the transmitted energy, leading to an overall accumulation of received energy.

It is the integration of these two factors (a flow) that allows us to evaluate the quantity of energy received, which can be interpreted as a risk affecting the nominal conditions required for a system to work properly (as per its service contract). Coupled with this receptor surface, let us imagine a hierarchical tree structure of detectors capable of determining the size of the reaction, or the necessary countermeasures to compensate for the risk incurred by the system (see Figure 2.5).

The nodes of the tree operate like an aggregate of organs with thresholds that send a particular signal to the level above in the structure when the sensitivity threshold is reached, and so on and so forth, down to the root, which, if it is reached, is associated with a panic signal. The greater the intensity of the flow, the higher the signal relayed by the scale of sensitivity will be.

We can imagine trees with a variable number of leaves depending on the sensitivity required; fewer leaves for a greater sensitivity, or many for the inverse scenario.

REMARK 2.3.– The retina is organized according to this principle, with an ultra-sensitive zone, rich in cells, called the fovea, where the retinal image is at its most precise. With stable trees, which have the same number of leaves per node, we find a classic logarithmic correspondence, depending on the basis for grouping leaves on the nodes.

We thus see the ease with which either nature, via arborescent networks like the thousands that exist in our nervous system, or human engineering, through its creative capacity, can create structures whose visible effect follows a power law adapted to the situation.

We can reasonably conjecture that behind every power law, there is an implicit hierarchical structure, or, on the contrary, an explicit structure such as the one found in the architecture of artificial systems. Herbert Simon sketched out the theory in his book, *Sciences of the Artificial* (see Chapter 8, "Architecture of Complexity").

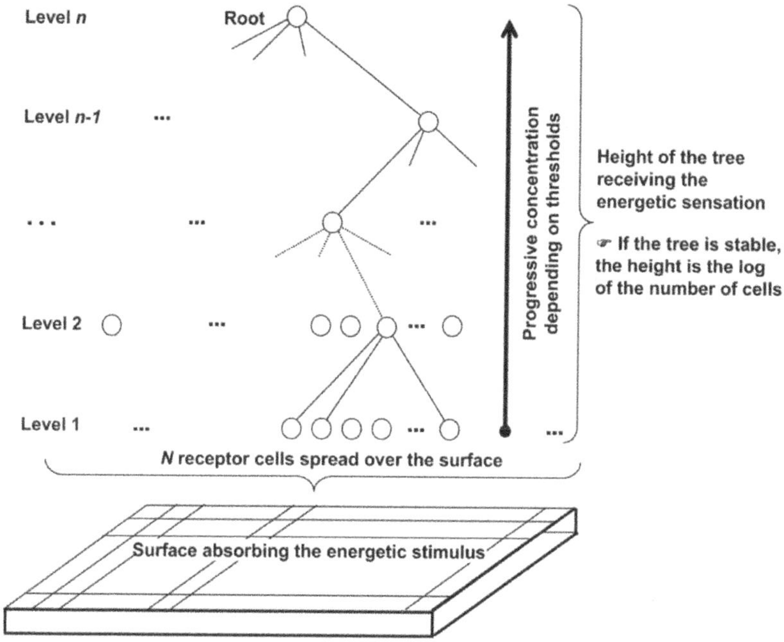

Figure 2.5. *Transducer with a power law*

Generalizing, to simplify the issue, we understand that the stimulation is tied to the space in which this stimulus appears: a line, a surface, a volume; in nature, it always occurs as volumes: a tube corresponds to a line and a plate to a surface (with one or two dimensions being negligible with respect to the third).

What causes the power law is the transformation of an energetic transfer on a surface or volume (2D or 3D), into a graduated scale along a line (1D), which will facilitate the "calculation" of the response $Stimuli \xrightarrow{Transduction} Response$ in the functional chain of the transducer.

One of the most fundamental structures created by nature and formalized in mathematics via the concept of a group is recognizable in this transformation. In this case, it is the reversible translation between two groups, group × and group +. In general, we can speak of transduction each time a correspondence is created between two functional spaces sharing one common, logical structure (the same "logical model" as von Neumann would have said), but incarnated as physically organic substrates that are specific to them, as in Figure 2.6.

The substrate spaces are physio-chemical systems from which we draw out abstract functional models that define the element/equipment of the corresponding system.

REMARK 2.4.– In an electrical system (see Chapter 7), we thus have two interacting systems (see Figure 7.8): a) a control system that processes information concerning the state of the electrical system, in other words an information system that exists physically on computing platforms whose programming component "materializes" the logical model; and b) the production–transport–consumption system of the energy produced, with up to 110 gigawatts of power over very real physical equipment, which contains a "virtualized" image in the information system that must remain absolutely coherent with reality in order to avoid serious disruptions.

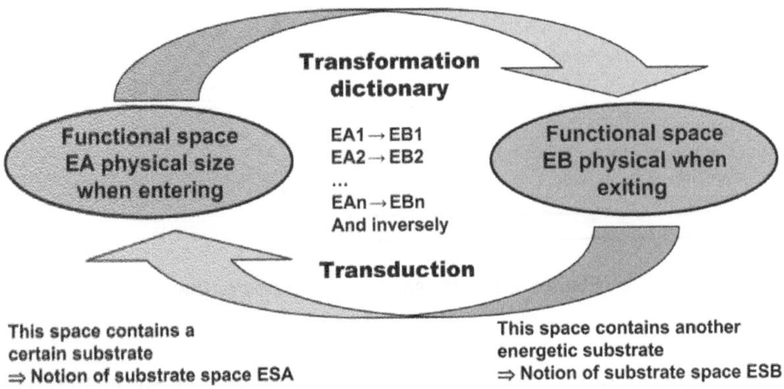

Figure 2.6. *Functional spaces and transduction between spaces*

One of the most interesting structures that has appeared with the complexification of equipment is the notion of the equipment's "grammar", understood as the set of rules for its usage that provide control over the equipment, with the understanding that the equipment monitors the elements of its own internal structure that only it knows (these are the MIB, or management information bases, in telecommunication equipment)[5]. This new structure, which emerged in the 1980s, made it possible to create a new generation of more elaborately "intelligent" transducers, providing more varied controls and options capable of accounting for the entire or part of the spatiotemporal context in which the equipment and/or system operates. The schematic diagram of these transducers, of their "logical design" as von Neumann would say, is shown in Figure 2.7.

5 See Printz (2013a, 2019).

A language is associated with each grammatical structure, which reduces the physical problem of energetic transduction to a problem of formal translation of logical structures founded on grammars, with the understanding that the organic part of the transducer has its own internal logic founded on energetic phenomena that are particular to it. This makes the equipment independent of other equipment, and frees the functional logic from organic contingencies while respecting the laws of physics.

REMARK 2.5.– We will see that in the case of the hierarchical stack of computing equipment whose base is made of integrated circuits at the nanometric scale, these are sensitive to the noise of α/alpha particles, which was not the case for analog and/or digital calculators of the first generation; this noise is compensated for by von Neumann's logical model of these first machines, a model that still persists in today's computers thanks to ad hoc redundancies (see Chapter 3, which addresses these stacks of computing interfaces).

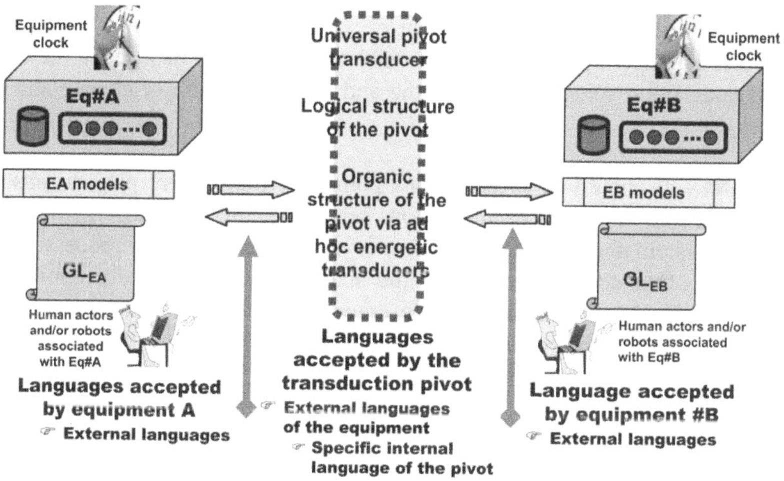

Figure 2.7. *"Intelligent" transducers*

This entire abstract "mechanics" is nothing more than a materialization of the Church–Turing thesis, which says that "anything that can be understood by the human brain and communicated to another can naturally be expressed in the language(s) of a certain Turing machine" yet to be discovered – by intuition and not by an algorithm – as in the "linguistic" approach suggested by A. Grothendieck[6]. In the "intelligent" transducer's architecture, whose structure we have just sketched out,

6 See his book, available on the Internet, *Récolte et semailles*. Grothendieck is a first-rate mathematician and recipient of the Field's medal; he has also taught in the United States. His book was just published this year with Gallimard.

everything relies on a) the Turing machines specific to each type of equipment, with the programmed part constituting a constructive "proof" of the thesis, and b) the Turing machine associated with the pivot that acts as a universal connector between different equipment. One of the roles of the system architect, surely the most fundamental for the mid- to long-term survival of such a system, is to discover this deep structure and to make a model of it that can be a) used by the equipment for processing the associated information and b) understood by those using it (customers and users/developers of the system, and even the stakeholders).

For the system and all the elements that constitute it to maintain coherence, the semantic constant that characterizes this coherence must be maintained, whatever the nature of the transformations being made, which implies a certain symmetry between the operations being performed. It is a symmetry that is also obligatory if we wish for the system to be reversible; that is, for any situation, nominal or non-nominal, we can reconstitute the chronology of transformations leading to the situation in order to verify/validate that everything went well. Whatever happens, the reverse engineering must remain possible, otherwise we will leave the confines of well-understood scientific engineering and enter the realm of chance and occult sciences.

IMPORTANT.– It should be clear by this point that transducers play an absolutely fundamental role in mediating interactions between a) the physical-organic, material part of the system and the synchronous or asynchronous energetic processes with or without mass that occur there, and b) the necessary information for controlling it, including in the human realm, and the ergonomic constraints of customers/users, which must never be forgotten.

2.2.1. *Example of an adder circuit with CMOS technology*

Adder circuits are part of a pool of hardwired functions that we have found in all the central units of computers in the von Neumann model, since the beginning with his electronic computing project (ECP) machine at the Institute for Advanced Studies at Princeton University. These circuits are transducers that synchronously transform an energetic state into a new state, which we can interpret as an addition. The electronic substrate of the machine, and the materials that constitute it, stabilize two physical states, denoted by 0 and 1. When a programmer writes a simple instruction in their favorite programming language such as $a = b + c$, they will end up activating an adder circuit implanted in the silicon of the microprocessor. It is the conception and the construction of this type of circuit that we are going to analyze, not to examine the details of the construction from the electronic perspective, but to understand how the creator formulates the problem of realizing a circuit using different languages that account for internal and/or external points of view, to create a circuit that in the end functions perfectly.

The reference works that describe how to go from CMOS transistors to different types of logical circuits are numerous; for example, among those which our logician colleagues used regularly in the 1980s–1990s: *Principles of CMOS VLSI Design – A System Perspective* by Weste and Eshraghian (Addison-Wesley), and *Introduction to VLSI Systems* by Mead and Conway (Addison-Wesley)[7]. More recently, the book by our colleague from the CNAM, François Anceau, *Conception des circuits VLSI – du composant au système* (Dunod, 2007), gives a good summary in French, and by someone who knows what he is talking about.

The first level of language used by "logicians", in the hardware sense of the term, is simply that of Boolean algebra. Ever since von Neumann, the internal language of the computer has been that of Boolean algebra, since physics and electronic engineers know how to make very stable (fundamental in engineering), two-state circuits, denoted by 0 and 1, and transform these states, via transduction, into another state, which is also made of 0 and 1, and which will constitute the result. The first step therefore consists of establishing correspondence tables for the operation ADD, as shown in Figure 2.8.

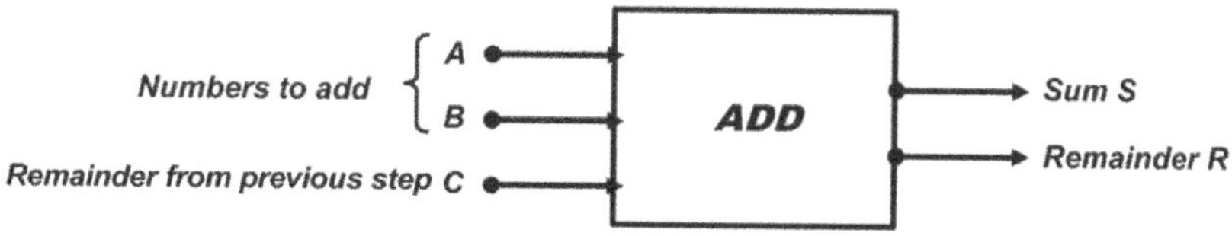

Figure 2.8. *Logical addition with remainder*

We will not give the details of truth tables – they can be easily found in any good reference work, including Wikipedia – since it is not what we wish to illustrate here[8]. We will simply remark that we must be careful with remainders, which must be correctly carried from one step to the next. In standard Boolean algebra, 1 + 1 = 10, which means that if A and B are equal to 1 (and there is no remainder to carry over), the sum *S* equals 0 and the remainder *R* equals 1. For the entry values A, B and C, there are therefore eight cases to consider.

7 These references were provided to me by my colleague at the CNAM, Georges Kéryvel, who lectures on operational research and was one of the designers of the ISIS super-calculator, under the supervision of Claude Timsit, and co-developed by Bull and the CEA.

8 See: https://en.wikipedia.org/wiki/Adder_(electronics).

Once the logical equations have been established, we can change the representation so that we can visualize the pathways of data through the base circuits that in the end constitute the ADD operator, as shown in Figure 2.9.

XOR is the function *OR exclusive* (either the one, or the other, but not both) and *NAND* is the negation of the traditional *AND* that we would write as *A AND B*, or even *NOT (A AND B),* with *NOT* being the usual unary operator for negation. The utility of this new representation is that it begins to resemble a real circuit drawn in 2D. We begin to see what happens and are able to represent the temporal sequence of elementary operations, another vital necessity in making a circuit that "works".

Based on this representation, we can begin to translate this kind of logic by introducing physical transducers this time, which will support the operation, in this case, CMOS transistors, whose technology became widely available in the 1980s.

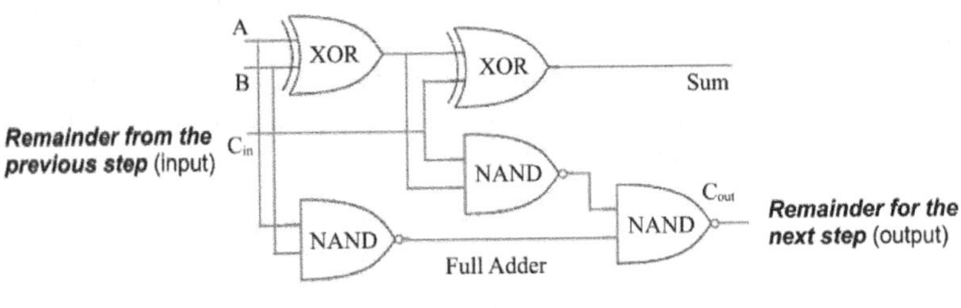

Figure 2.9. *Diagram of logical addition with a remainder*

CMOS transistors work like simple interrupters, are easy to make, requiring little energy, and above all, are easy to make in miniature. Today, after 30 years of continuous R&D, they are of nanometric size, integrated into systems called SOCs (System On Chip) that can include many millions of transistors (Figure 2.10).

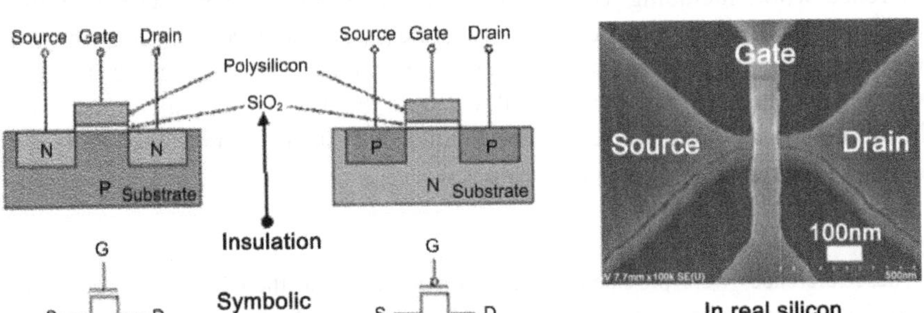

Figure 2.10. *CMOS transistors*

The figure shows the two types of CMOS transistors and their symbolic representations, as well as a cut into real silicon where we can see the precision of the engraving. Here still, the issue at hand is not to explain transistors, but to show how we can go from one representation to another, until we arrive at the adder circuit with 28 transistors, as shown in Figure 2.11; this is only an example.

Starting with this kind of elementary circuit, we can, depending on the performance constraints, create or iterate the basic geometric pattern in order to add 16, 32 or 64 bits depending on the need. If we have enough transistors, we can preferably choose the composition of circuits by duplicating the pattern as many times as necessary. On the contrary, if we must save the resource of transistors, the creator can choose an iteration as indicated in Figure 2.12a. It gathers three of the representations we have spoken about: truth tables, a diagram with XOR and NAND ports, and an iteration sketch that we could represent in design languages such as VHDL, as well as the classic representation of robots, denoted by (b) in the figure.

Passing from one pared-down, iterated representation to a higher-performing, composite representation is the same as "unfolding" the iteration cycle, which is a common transformation operation in compilation techniques that can be done with total confidence. On historical computers such as the Whirlwind I and II, the iterative solution was adopted. In modern computers where there is an abundance of transistors, the duplicative solution is used, keeping in mind that for an adder of 32 bits, at least 32 x 28 = 896 transistors will be needed and a great deal of cables to connect them to each other, which is another problem that it is possible to control.

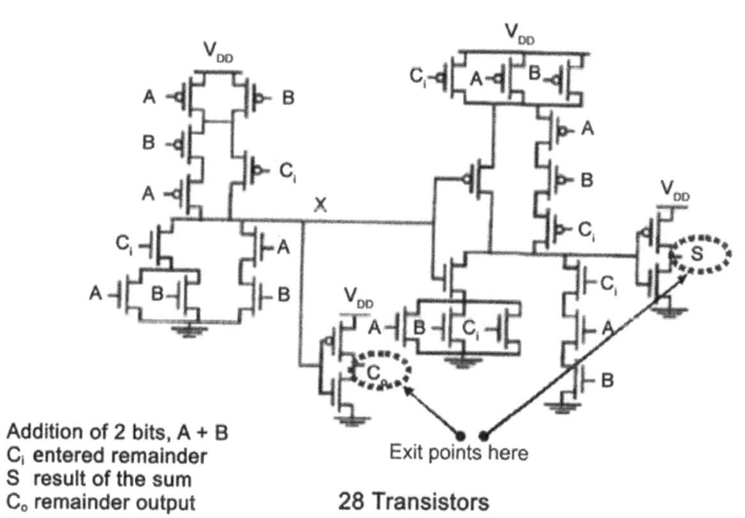

Figure 2.11. *Adder circuit with CMOS transistors*

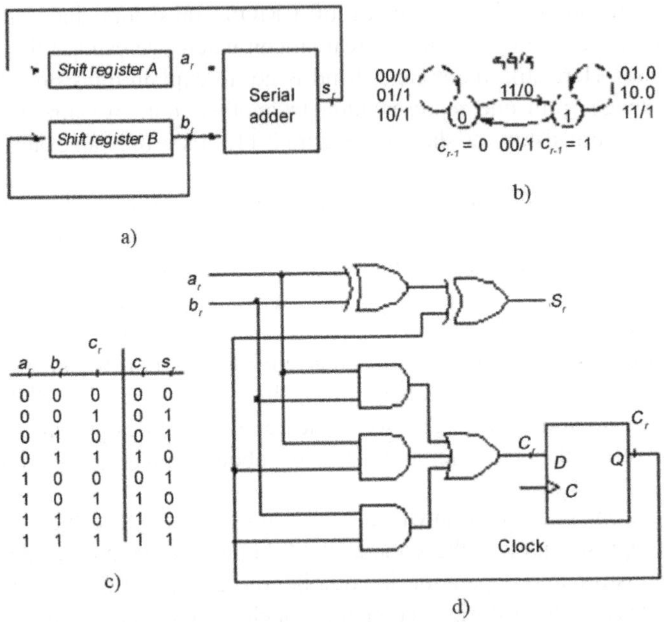

a_r	b_r	c_r	c_r	s_r
0	0	0	0	0
0	0	1	0	1
0	1	0	0	1
0	1	1	1	0
1	0	0	0	1
1	0	1	1	0
1	1	0	1	0
1	1	1	1	1

Figure 2.12. *Iterative adder circuit*

If we wanted to be exhaustive, we would have to discuss the language used to define the "masks" that allow us to control the engraving machines that implant the transistors and the wire conductors that connect them to each other on the silicon substrate, as indicated in the previous figures (see also Figure 2.13).

For all of the aspects tied to CMOS arithmetic and the way of creating elementary circuits to obtain arithmetic functions over 16, 32 or 64 bits, as well as the "masks" required to install the wiring, the book by Kai Hwang, *Computer Arithmetic: Principles, Architecture and Design*, has for a long time been the canonical reference for engineers and logicians.

The wiring of a VLSI is a highly complex operation, which von Neumann would certainly have qualified as "ultra-high complexity" in the terminology he used in his Hixon lectures. It caused suffering to all those who encountered this revolutionary and almost "miraculous" technology for an observer outside of the field. To connect one point to another on a silicon substrate, we connect them according to the axes presented in the drawing in Figure 2.13.

To go from A to B, the smelters always work, for technical reasons, along the two axes, horizontal and vertical, on two different planes to isolate the connectors from each other, because crossing paths without meeting requires at least two planes. To go from A to B, a "hole" *H* must be made to simply change the plane, the distance being $l1 + l2$. Knowing the precise distances is indispensable for determining the performance of the circuit and the switch times.

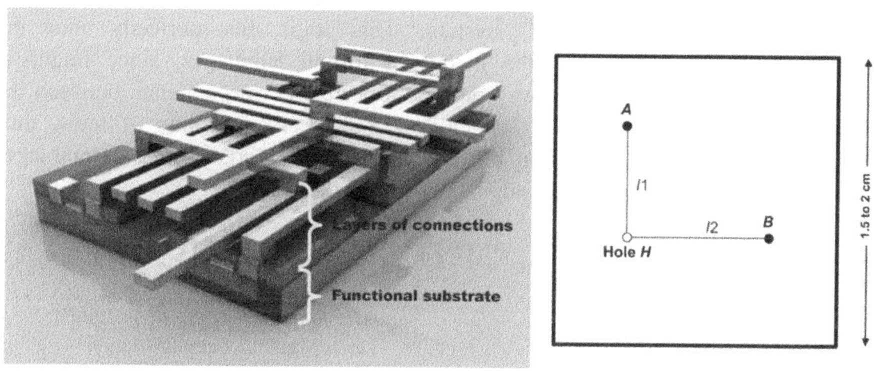

a) b)

Figure 2.13. *Connection elements*

The artist's rendering (Figure 2.13a) shows the principle of these connections, which in current technologies can be made over a dozen or so layers, depending on the number of components to be connected and the physical constraints tied to the materials being used.

When there are 1 or 2 billion transistors in the circuit, the sums $l1 + l2$ end up creating distances that can be counted in tens of kilometers, and more than two planes are required to organize the weaving given all the constraints (usually a manufacturer's secret), which makes it impossible to reverse engineer the weave, and consequently to reconstruct the logic of the circuit by examining the silicon. This is very useful for protecting expertise, but it is also an enormous problem for validation since we can no longer test things "by hand": they are too small, and moreover, too delicate. It is here that the theory of languages and compilation techniques will show its strength since the engraving machines are piloted by the language of successive masks (a standardized interface, specific to the smelter) that will simultaneously allow tests to be generated, guaranteeing, if all goes well, that the weave is correct in terms of electrics.

The moral of the story is that we see nothing: it is too small and too delicate, but we can nevertheless exhaustively validate it using tests that the language of masks lets us generate.

This aspect of modeling languages gets used systematically, including for the functional logic of the circuit, either with representations using transistors, or with diagrams or logical equations, and all these linguistic texts have the final objective of coherence across the SOC system. This logic fits perfectly into the methodological framework of the theory of formal languages, since linguistic abstractions constructed in this way provide a perfect correspondence between the abstract and the real, modeled from all of the necessary points of view, thus providing a semantic duality that lets us exhaustively validate these assemblages built upon silicon, which, in the end, make perfect transducers.

3

Emergent Properties: System Integration

This chapter begins to illustrate some of the fundamental concepts summarized in Chapter 1, using the example of one of the most emblematic objects which each of us has access to 24 hours per day, 7 days per week, in real time, to interact with our environment, wherever we may be (or almost wherever) on the planet. The object is the smartphone and its stack of interfaces, which includes everything from the molded silicon used to engrave the integrated VLSI circuits, today called the SOC (System on Chip), to the purely immaterial interfaces which provide access to social networks and all sorts of applications and services that are useful for interacting in our complex societies where "everything is connected."

3.1. Integration

The user of a smartphone or tablet, thanks to which they can interact in real time with a multitude of systems from across the planet, probably does not imagine that what they hold in their hands is the quintessential scientific and technological object of the 20th century, and probably the most elaborate example of the principle established by the physicist Philip Warren Anderson, a master of condensed matter, the modern name of the physics of solids (or solid state physics). It is the science of the disorder of materials, which nature provides in abundance and physicists and engineers try to give order to so that properties that would not exist without these improvements can emerge; and so it goes for the modified silicon of our computers[1]!

For a color version of all the figures in this chapter, see www.iste.co.uk/printz/pedagogy.zip.

1 See *La Matière en désordre* by E. Guyon et al., already cited.

In 1972, P.W. Anderson (born in 1923, won a Nobel Prize in 1977, and had a prestigious career at Bell Labs, Cambridge and Princeton), published an article that would become famous: "More is different: Broken symmetry and the nature of the hierarchical structure of science"[2]. It detailed a series of observations, inspired by his experience, that in a way completely ruined the reductionist philosophical stance, but with some precautions, and launched the basis of integration science and what would come to be known as the physics of emergence. Without this mastery of integration at every level of the computer stack, as we will see, computers would simply not exist! He said:

> The ability to reduce everything to simple fundamental laws does not imply the ability to start from those laws and reconstruct the universe. In fact, the more the elementary particle physicists tell us about the nature of the fundamental laws, the less relevance they seem to have to the very real problems of the rest of science, much less to those of society. The constructionist hypothesis breaks down when confronted with the twin difficulties of scale and complexity [...] at each level of complexity entirely new properties appear, and the understanding of the new behaviours require research which I think is as fundamental in its nature as any other [...] The arrogance (say, of some) [...] may be behind us. [...] At each stage entirely new laws, concepts, and generalizations are necessary, requiring inspiration and creativity. [...] Surely there are more levels of organization between human ethology and DNA than there are between DNA and quantum electrodynamics, and each level can require a whole new conceptual structure.

Information sciences, which were developed by von Neumann, Shannon, Turing, Wiener and a few others, produced one of these new conceptual structures: the computer, or von Neumann's "computing instrument" which existed in a double form: a) as a theoretical abstraction, a robot with memory that would later be called the Turing machine in homage to Turing for demonstrating the mathematical properties specific to this new science, and b) a functional model of a real machine, von Neumann's architecture, which he himself called a "logical design" for the engineers who created these "miraculous" machines, a model which is still in use ever since its invention in the 1940s.

---

2 *Science*, 177(4047), 393–396.

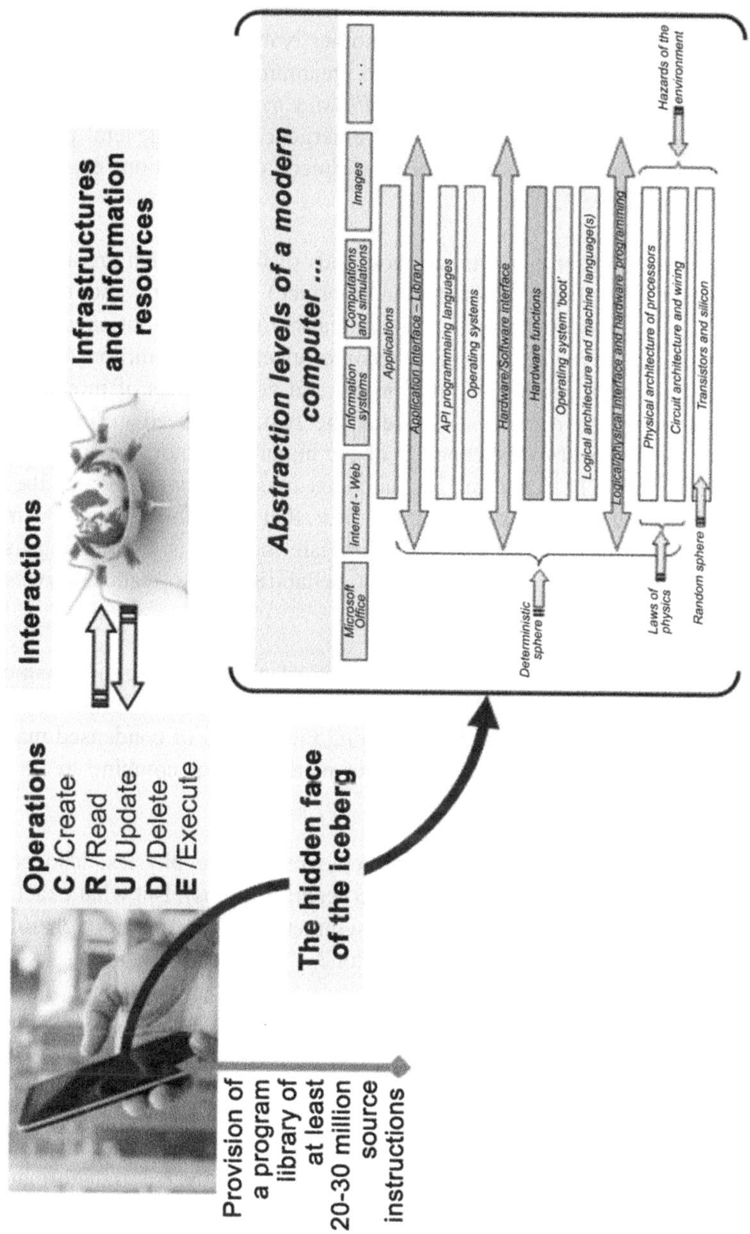

Figure 3.1. *The computer stack in its user context*

The computer stack, present in every computer and handled billions of times by users every day, is the best illustration of the principle of "more is different," as Anderson's colleague Robert Laughlin, another Nobel Prize winner in 1996 and a professor at Stanford, reminds us in "The Quantum Computer", Chapter 6 of his book *A Different Universe – Reinventing Physics from the Bottom Down*, in which he provides an account of this physics of emergence for the general public. In the hands of this "Thumbelina," it was a masterpiece of abstraction, engineering and integration (see Figure 3.1)[3].

The actions of a user can activate thousands of functions that are accessible via the equipment's HMI (human–machine interface). The entry and restitution of information, from *outside* the system, is done at human speed, but once it is entered into the machine, the *inside*, the speed becomes electronic, at millions of operations per second for the processing of information and at thousands of interactions with the exterior world via networks. Between the two, the speed differential is 1 to 10,000, but with respect to the capacity of the human operator, we are closer to the order of a million! The stack and its functions must otherwise respect the service contract introduced by the use that is made of it, that is, by the general constraints of the environment (the *outside*) that we abbreviate in systems engineering using the acronym FURPSE: functionality, usability, reliability, performance, serviceability, evolution[4].

But before detailing the structure of this stack and presenting some of its fundamental features, let us say a few words about the notion of "broken symmetry" to elucidate what Anderson meant concerning the physics of condensed matter, that is, the physics of integration of composite materials that combine to form much larger structures with new properties.

"Broken symmetry" is a recurrent theme in particle physics, also called the physics of high energy (much is required to separate atoms). But what exactly do we understand by the expression "broken symmetry"? What is its relationship to energy? And better yet, to systems integration?

Under certain energetic conditions (of temperature or pressure, for example), matter presents characteristics of reversibility or continuity that make its stable form invariable. Any modification can be compensated for by an inverse modification, thus the expression of "symmetry," but the compensation that occurs is formally closer to the structure of a group in which any transformation T that respects the rules can be cancelled and compensated for by an inverse transformation that

3 See the book by Michel Serres, by the same name, about his granddaughter nicknamed "Thumbelina" or *Petite Poucette* in French.

4 Norm ISO9126, as already indicated.

reestablishes the equilibrium, which we will denote by $T \oplus T^{1} \rightarrow e$, with e being the neutral element of the group and $\oplus$ the transformation operator. We can always take a step back and undo what was done, which is a fundamental requirement of engineering. We know for example the importance of the notion of the group in the construction of geometric space[5].

A liquid that gets heated or cools off remains liquid, expanding and contracting slightly, but conserving its liquid state (a collective state of the particles, so that in this state, the liquid occupies the volume containing it, in other words, the geometric form of this volume), as long as the boiling point and the freezing point are not exceeded, which would transform it into a gas or a solid.

A straight beam that is compressed or stretched by pressure/traction (see Figure 3.2) maintains its straight form up to a certain point. Beyond this point, the beam bends, which means that the pressure is now compensated for by two parameters: contraction and flexion, and perhaps torsion.

In these two cases however, beyond a certain energetic threshold (heat, elastic compression), the configuration of the system changes radically; the water evaporates, becomes a gas and loses its properties of water (an event which is feared in the furnace of a steam engine). There is a bifurcation or even a singularity, in the mathematical sense of the term. It is this type of phenomenon that we call broken symmetry, "broken" in the sense that another property, that was not in the initial system, appears, or on the contrary, that an intrinsic property disappears suddenly, given the conditions of {T, P}. From a systems engineering point of view, it is imperative to fully know the modalities of this kind of rupture in order to prevent them; it is this worrying aspect that is at the heart of what is called functional assurance, whose objective is to maintain the system in a nominal state as defined by its service contract.

We should point out that in the case of a system object made of a beam (see Figure 3.2), the length varies very little due to compression and/or thermal expansion/contraction, and we can anticipate the wiggle room that must be given to the structure. But when the bend appears, there is discontinuity (it is a singularity) due to the presence of the bend itself, which can provoke a complete destabilization of the structure in the case of a mechanical assemblage. This is the phenomenon called buckling, exceedingly dangerous, and a prelude to certain catastrophe if the deformation is not rapidly compensated for.

5 In addition to what Poincaré had to say about this in *Science and Hypothesis*, see the works by Jean Piaget on cognitive psychology.

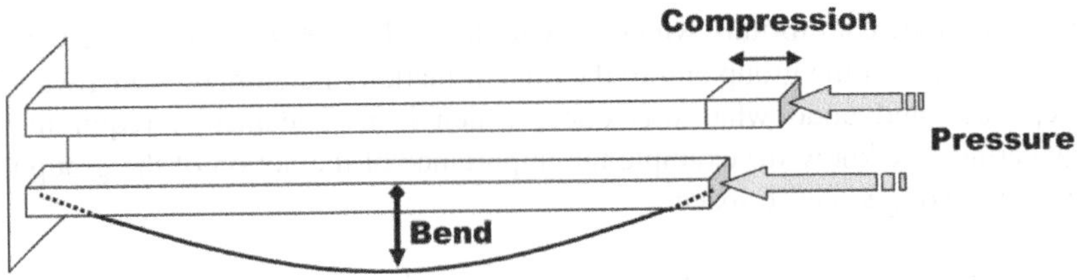

Figure 3.2. *Broken symmetry: beam compression*

The energy transferred to the system through the exerted pressure can therefore be "digested" in different ways, with the possibility of a change in form: that is, a singularity. In the case of a beam, phenomena of bending and torsion will appear, possibly giving rise to a spiral form, like a helicoidal spring.

In nature, there are, from entity to entity, certain equilibria that depend on the energetic state of the body in question. An input of energy can radically change these equilibria, as in the case of tsunamis, where a sudden change in the depths of the sea liberates an amount of energy that must obligatorily be compensated for in order to respect the great phenomenological laws (conservation of energy in the case of a tsunami, which causes gigantic waves moving towards the coast to compensate for the lack of depth).

A gas does not have a form; it occupies the entire space that it is given; we can therefore compress it with ease, which can be a way of storing energy to the point of liquefaction, the critical moment when it loses its gaseous properties and becomes incompressible.

A liquid occupies the form of the space that contains it, but the cohesion of its molecules makes it incompressible (or virtually so); it transmits pressure, as Pascal showed in the barrel experiment. It is for this reason that hydraulic circuits were often used in the first servomechanisms; with ad hoc pipework, pressure – or rather, energy – can be transferred to where it is needed.

A solid is characterized by its geometric form, which keeps its shape but can undergo different deformations (each with a different threshold) that are all ways of storing energy, following three parameters at every point of the solid: contraction/stretching (a tangential property of the first order), flexion (bending) and torsion (the torsion radius of double curvature), the last two parameters being of the second order; hence the use of tensorial calculations to study deformations and in a general sense, all energetic phenomena.

These different qualities (or virtues, in the Aristotelian sense) are collective properties, for a particle alone does not possess any of them. Temperature, for example, is a collective property of a group of particles, given that the temperature of a particular particle is a notion that has no meaning.

The metaphor of the computer stack, which will be explained in detail in the rest of this chapter, shows that analogous mechanisms exist in the abstract "computing matter" that give shape to this stack, with the understanding that the corresponding physics is yet to be established, even if we are beginning to have a good sense of it (see the notions from systemics recalled in Chapter 1).

REMARK 3.1.– The remaining question is to know whether such phenomena exist in the "physics" of social groups, as sociology was first called, given that what we know about the behaviors of these groups over the short, medium and long terms from their interactions (or from human ethology, to cite the term used by Anderson in the previously cited text). We will provide an example in the case study presented in Chapter 8, "Systemic Aspects of a Project System," which examines socio-dynamic phenomena that all project managers know well, even if they do not conceptualize them.

The foundation of architecture in layers is the notion of recursiveness as it appears in the theory of calculability proposed by Turing himself. The concept of the Universal Turing Machine (UTM) lets us represent any Turing machine (TM) specific to a particular problem in terms of the UTM. For each layer of the architecture, there is a corresponding TM that is adapted to the organic context of the layer, with ad hoc programming, which provides perfect control over what happens in the layer and allows for interaction with the above layer via a new functional interface in which the logic of this new layer can be expressed, given the specific constraints of the new layer.

However, in systems engineering, in practice, it is the notion of the real abstract machine, von Neumann's "logical design," that gets used to functionally describe what should be done given the proposed construction materials, in this case computer programming languages.

We can in this way construct a logical, rigorous edifice with guaranteed properties for a given environment. By taking a step back, we can assimilate this construction to the establishment of a code that corrects random errors, that is, the background noise generated by the environment the system is made in, following the same logic as that used by Shannon in demonstrating his second theorem. If we have enough information about the "noise" of the environment, then we can develop an architecture that compensates for this noise and make it so the system, and the construction procedures that are part of it, complete the task assigned to them; in other words, for any transformation T that occurs, we know the inverse T^{-1} so it can be repaired. The knowledge of this noise is therefore essential, from a project point of view, and otherwise, we must be able to observe its behavior so that any deviations can be detected and corrected (this is the autonomic cycle of systems engineering)[6]. For this kind of observation to be possible, the functional assurance model implanted in the system must be formalized so that for every difference, we can see what will be useful in making future corrections. It is an application of the fundamental principle of systems engineering and systemics, which demands a clearly defined boundary between the inside and outside, even if only conventional, as recalled in Chapter 1.

3.2. The stack structure and its constructive logic

After an evolution that began with the first computers, the structure of this stack can today be summarized as in Figure 3.3.

At the base of the stack, we find silicon monocrystals, subjected to the rules of the quantum realm and different kinds of thermic and electromagnetic "noise," α/alpha particles coming from natural radioactivity and the cosmic rays that cover the surface of our environment on Earth.

The consequence is that a transistor, nanometric today thanks to the progress of the CMOS which has allowed for its miniaturization (see above), can change energetic states randomly, making the processing of information memorized in this form just as random, if nothing is done about it. The correction codes for the error (ECC, or Error Correcting Code) that act at the level of the circuits, as depicted in the figure, act to compensate for these induced errors; in other words, they: 1) detect and 2) correct (after simple malfunctions, suggesting that there is only one transistor in the monitored zone). Without these codes, a modern computer would not work correctly for more than a few minutes, given the number of transistors it contains and the preponderance of rays that engulf our world.

As soon as we reach the level of "logical architecture and computer languages", the computer becomes entirely deterministic – or Newtonian, we could say – and it can be programmed; in other words, it will follow all the commands of hardware programmers reliably and exactly, but only at this level. It remains unprogrammable by software programmers, in the usual sense of the term.

What is remarkable in this stack is that the same principles of organization, "generalizing to simplify" among others, apply from one level to the next, from the

6 See Printz (2019, Chapter 4).

base to the summit, which to some means that the architecture becomes "fractal" – an interesting and vivid image, to be sure, but that does not account for the logic of construction and does not allow for concrete action via the engineering of interfaces, the *how* (see Figures 3.10 and 3.11).

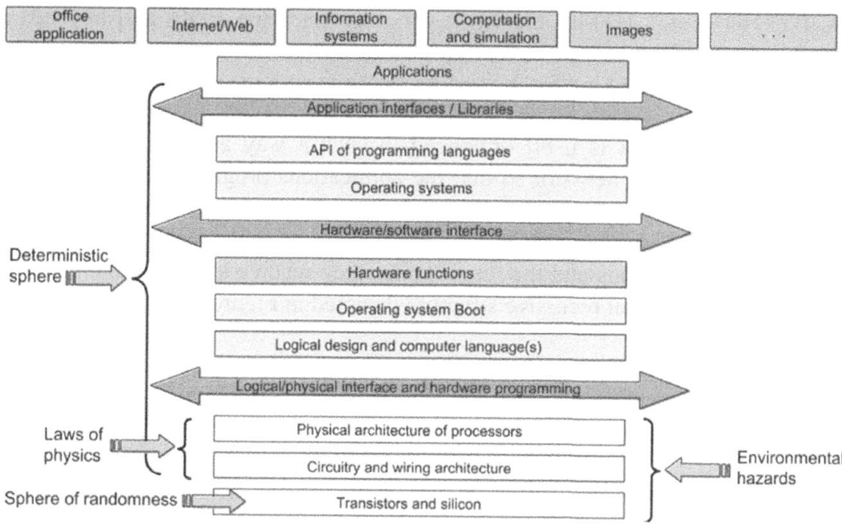

Figure 3.3. *Structure of the computer stack*

To make the computer user-friendly, that is, so it makes the user forget its organic structure through its external language, different screens are needed, such as operating systems and network stacks for external telecommunications and access to networks, the medium for interactions between equipment and the rest of the world. A similar apparatus also exists for disk input/output and permanent memory (access to databases, database management systems (DBMS), file systems (FS), network file systems (NFS), etc.), not represented here for the sake of simplicity.

At the hardware/software interface level, the computer becomes programmable for systems programmers who make operating systems, telecoms, graphic interfaces, etc. using integrator languages that exactly reflect the structures of the computer that the systems programmer seeks to pilot. Very early on in the history of computing, integrators were complemented by so-called "high level" languages that facilitated the task of systems programmers, for example, C (a simplified version of the PL1 language invented by IBM in the 1960s). At this level, the programmer needed to know how a computer works and be familiar with the libraries of available systems programs (several hundreds of functions via API systems require two to three years of learning to master the content). A video game or control/command systems programmer must master this level, although it is too complicated for applications programmers, who constitute the majority of programmers (90 to 95%).

For the highest level, the top of the stack, the applications programmer has access to languages (FORTRAN and COBOL since the 1960s, then C/C++, Java, C#, PYTHON, etc.; HTML and XML for data, since the 1990s and beyond) and libraries that let them "forget" everything below so that they can concentrate on the application they are making for the end-users of the system.

The network stack is itself organized in such a way as to mask the intrinsic non-determinism of a network so that the applications programmer may efficiently control what they are programming and the effects it produces in the system environment. The programmer has no knowledge of what really happens except through the notifications and the return codes they receive after entering code. This is what the fundamental recursive structure depicted in Figure 3.3 shows.

Today, with engravings as small as 20 nanometers, or a row of 150 to 200 atoms, a transistor is a small volume that contains, in 3D, somewhere in the order of 4 to 8 million atoms acting as a collective whole.

REMARK 3.2.– Let us recall that the Avogadro number is approximately 6×10^{23} atoms per 28 grams of silicon, for example, or 12 grams of carbon; its crystalline structure is very close to that of a diamond (a carbon crystal; both elements in the same column of the Periodic Table). The hazards and fluctuations of certain individual atoms are compensated for by others who are not affected at the same moment, similar to what happens with crowds.

It is as a group that all these atoms are able to create the desired, collective effect, materialized by a perfectly controlled and above all stable, final energetic state (robust and resilient). We could not say what the threshold is for perturbing the collective effect with quantum hazards today, but we are certain the threshold exists, as indicated in the figure. At this stage, it is a haphazard endeavor since the calculations produced by quantum mechanical equations are inapplicable because they are too complex (with N particles, there are already $6 \times N$ degrees of freedom and as many equations needed to solve the system).

The process of making a circuit that contains a few billion transistors uses many hundreds of engraving and verification machines, which means that between the entry of a silicon wafer at the start of the production line until the moment a useable circuit for systems tests is obtained at the end of the line, the length of the fabrication process can take somewhere between six and nine months, depending on the complexity of the circuit. Which means that, from an engineering point of view, any error, just one error in the design, or perhaps a few minutes of distraction, will mechanically lead to six to nine months of delay in the project. This is good example of broken symmetry in which a punctual event causes a vast overall effect.

In order to connect the transistors to each other (so that they produce the expected functions) and provide them with electric energy, a chip may contain hundreds of kilometers of nanometric wires, invisible to the eye and a light microscope, cabling that may itself contain certain defects.

At this stage, we can talk about a certain collective infallibility, given that individually, a mistake remains probable, but it never simultaneously affects every member of a community (see the paradox of birth dates in the study of how reliable memory is when a memory prompt *and* its corrector code are simultaneously affected by two almost synchronous particles, in other words an interaction whose duration is inferior to the time it takes a clock to complete a cycle; it is a probability that we know how to calculate).

To fully understand constructive logic and the complexity that an architect of such systems must organize, we must never forget the fundamentals of any construction containing the triplet that nature offers us: matter, energy and information (in this case, any programming/organizational component, which nowadays has an enormous role in systems; see Figure 3.4).

Every system is made of materials. In the case of computers and ICT (information and communications technology) in general, the flagship material is silicon, one of the most abundant elements on Earth. For the system to function correctly, the properties expected of the materials must be guaranteed, taking environmental constraints into account, hence the stage of "tests and diagnostics" systematically associated to each of the functional levels. For the silicon and other materials present in the chips and system cards to keep their properties, the temperature must not go beyond certain values, nor must they be exposed to destructive rays; beyond the range of validity, the material is no longer in the nominal state guaranteeing its proper functioning. The system must imperatively be stopped.

The electrical level enables the transformation of "formless" energy, which arrives via the electric power supply, into a useful energy that is "formed" through the informational logic of the system's organization. A modern chip made of billions of integrated transistors and electronic components is initially a gigantic network of hundreds of kilometers of electric conductors made of copper or aluminum, sometimes gold, of nanometric size, and which all conduct at the speed of 1 billion cycles per second. Applying the Ohm and Kirchhoff laws, "things get heated" and the heat must be dissipated at the same rhythm as the one at which it appears. If there are short circuits, parasitic capacities, currents induced by the environment or unforeseen phenomena, the circuit will not work. If the constitutive hardware material has defects, we will not be able to say, in the case of a circuit failure, if the breakdown is due to the circuit itself or to the hardware material. A diagnostic becomes impossible.

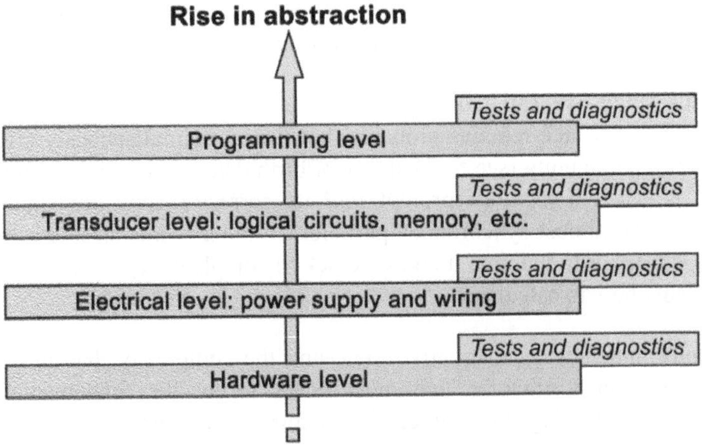

Figure 3.4. *Constructive logic of matter in circuitry*

This same logic is repeated from level to level, each time with specificities specific to that level. At the "transducer" level, the proper functioning of the adder circuit or the reading/writing of a memory cell or register presupposes that the underlying levels function correctly. The "tests and diagnostics" level is specific to the level in question. To satisfy its function, it must not consider what happens beneath nor the use that is made of it above; for that we say that it is independent of the context, or context-free in computing jargon.

As for the programming stage, which is the real development brought about by the computer and its logical design, it is made using external languages, whether explicit or implicit, in the sense that lower levels provide a view into their mode of functioning. This is a fundamental notion, in the strongest sense of the term, for it lets the programmer activate the functions they need at the level where they are working, without having to know the physical-organic details of the transducers that will enact the real energetic operation. The external language defines the nature of the observations and operations that can be made at this level and enables functional validation, hence the expression "design to test," which indeed captures the state of mind that systems architects need to have in order to define it correctly, in other words, to make sure it is testable.

3.2.1. *Correctly interpreting events*

An anomaly in the way a level works will happen at some point in the form of one or more events appearing at this level. Figure 3.5 summarizes and completes the two previous figures by presenting the perspective of the program/programmer. Seen by users and application programmers, the system outlined in red dots is, in theory, a perfect "black box" whose internal mechanics (i.e. the internal language and the associated transducers, defined by its grammar) remain unknown to them. This means, concretely, that all events affecting the function of the black box must be traced back to the inside/outside boundary materialized by professional application programming interfaces (API) and translated into the language of the profession(s) or the applications that are on "the outside". This is what we call a trace.

Let us give an example of a frequent situation with networks. When a network does not work or works badly, its monitoring system generates events called time-outs, among others, which means that the network no longer responds within the time frame that is normally allocated to it. The event is materialized in a message sent or expected by the application, and this bears no relationship with the content of the message. If the message is important or has priority, the application must perhaps be closed and saving procedures must be initialized; but if this is not the case, the error can be judged as benign, which means it is possible to carry on, having nevertheless kept a trace of the event. A network simply cannot respond after an overload of messages waiting to be processed; in this case waiting is enough. If there is a physical rupture, the network must be reconfigured, which necessarily takes a certain amount of time.

In the case of an anomaly, the black box must be observable at the interface because this allows us to interpret what is happening in the professional language employed by the user of the level. If that is not the case, that is, if the inside/outside boundary is porous, badly made or inexistent, we will have a profusion of useless data (see the Deceitful turkey effect, vividly named by Laughlin, Nobel prize winner, who explained how a profusion of data can completely derail research into the meaning of this data in the absence of an interpretive model)[7]. This applies to all levels of the hierarchy since at each level, there is an associated interface which can either be made using an explicit language or not (for the interface language, see Figure 3.10).

7 The deceitful turkey refers to a strange behavior among certain birds, observed by ethologists, to direct predators away from nests. For details, see Chapter 12 (p. 149), "The dark side of protection," in Laughlin's book, *A Different Universe*. What Laughlin explains in this important book formally contradicts the assertions of so-called experts of big data who claim that data is self-sufficient without worrying about how it is collected.

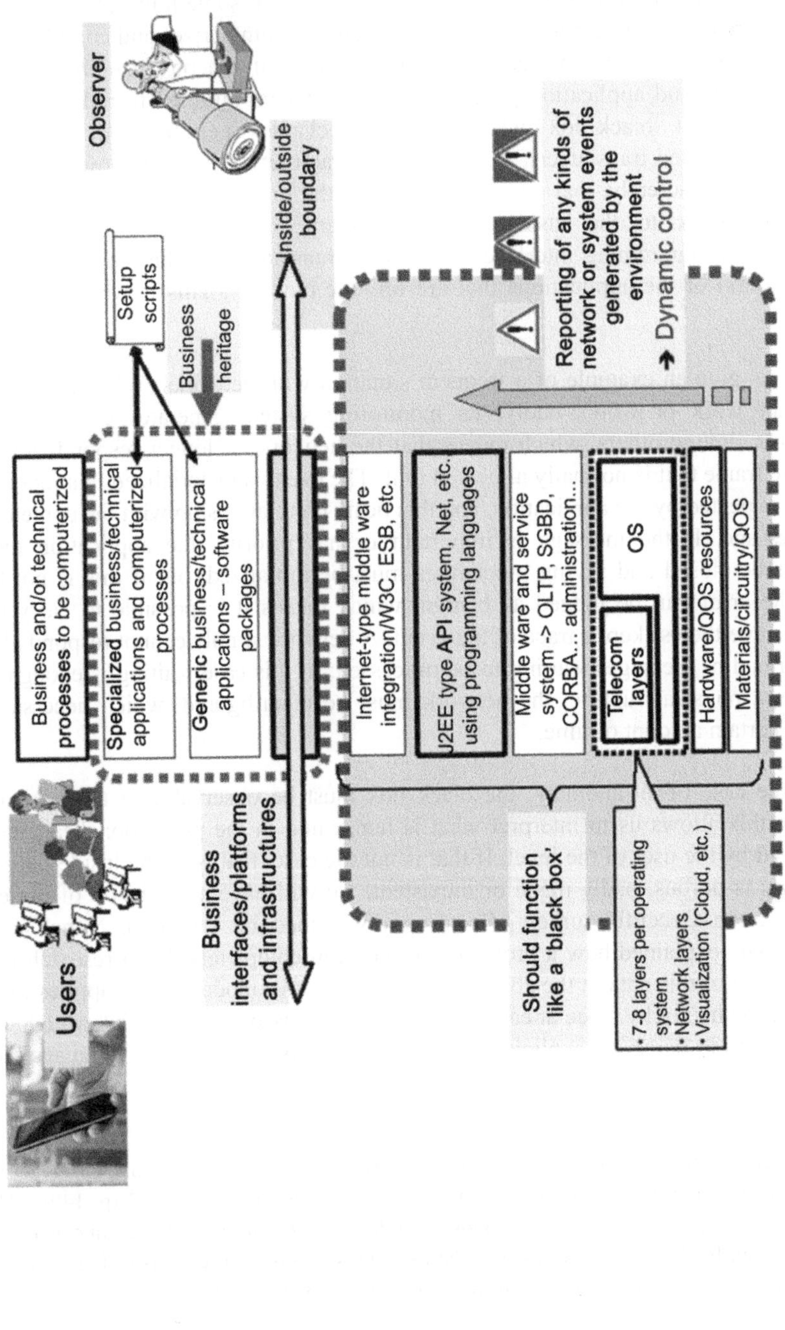

Figure 3.5. *Organizing the circulation of events in the stack*

At the “professional” level at the top of the hierarchy, we find, for organizational reasons, a hierarchical complement whose interfaces are purely informational, following the logical design as explained before (see Figure 3.6).

At this level of abstraction, there is no specific equipment beyond the connections ensuring communication between systems.

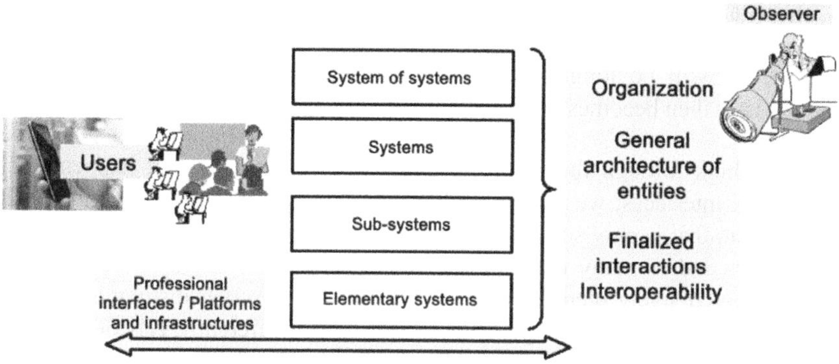

Figure 3.6. *System and sub-system levels*

The hierarchical systemic vision of systems → systems → sub-systems → elementary systems → equipment → functional modules is classic in systems engineering[8].

3.2.2. *The epistemological lesson of the computer stack*

There is a great lesson to be learned from all of these mechanisms.

If the observer turns their observation instrument to a direction other than the inside/outside interface, they will “see” many things that will not help them to understand what is happening, and will perhaps even send them down misleading paths, as with Laughlin’s deceitful turkey effect. We have existential proof of this in computing systems, which is the case for all systems nowadays. At the very least, the function of computing is precisely the ability to observe what is happening physically at the level of transducers, and this in an abstract way, via the languages

8 See the *NASA Systems Engineering Handbook* or the *INCOSE Systems Engineering Handbook*, which can be easily downloaded.

provided by the transducers (as is the case of management information bases (MIB) in network equipment, but this is only one example).

The logic of emergence imposes a logic of inside/outside which must be built by the creator, as suggested in Figures 3.7 and 3.8, with each layer being accessible through a library that is specific to the layer.

This is a necessary condition to the observability ensuring that gathered data has a meaning; the creator knows exactly what they are for. If the levels are not kept separate, we will no longer know where information comes from, nor what its function is; it then becomes impossible to understand what it means.

If we think through the complete pathway of everything that happens on the hierarchical interfaces, we end up with a very large amount of information overall (for a system that is only somewhat complex, this can potentially amount to billions of measures and as many messages and/or events) that make up a nesting doll of languages, recursively blended together without the observer distinguishing where the inside/outside boundaries get crossed among the different levels of abstraction. For example, the processes tied to error correction codes (ECC), which by definition are never the same since they depend on the hazards of the environment, from one execution to the next, give an impression of chaotic randomness even though they are anything but.

Inversely, if the system is only observed at the level of superstructures (if the instruments are not detailed enough, or if the instruments are pointed at the wrong places), we will never see the work that is done by the ECC, and we will never understand how the system acquires the fundamental property of determinism, which will then seem like the "miraculous" product of hazard's magic, as some biologists would say.

Von Neumann's wise recommendation (and also Descartes') follows on from this: always go from the simplest level towards increasing complexity, rigorously controlling every stage of integration, and never the reverse.

These are the ABCs of constructivism done right, hence the fundamental methodological role of always reasoning with explicit models, even if it means changing models while managing changes, but this is not so easy to do since we are juggling semantic abstractions, which are the most difficult to see without serious training. It is even one of the principal difficulties in systems engineering pedagogy, especially when it is learned through books.

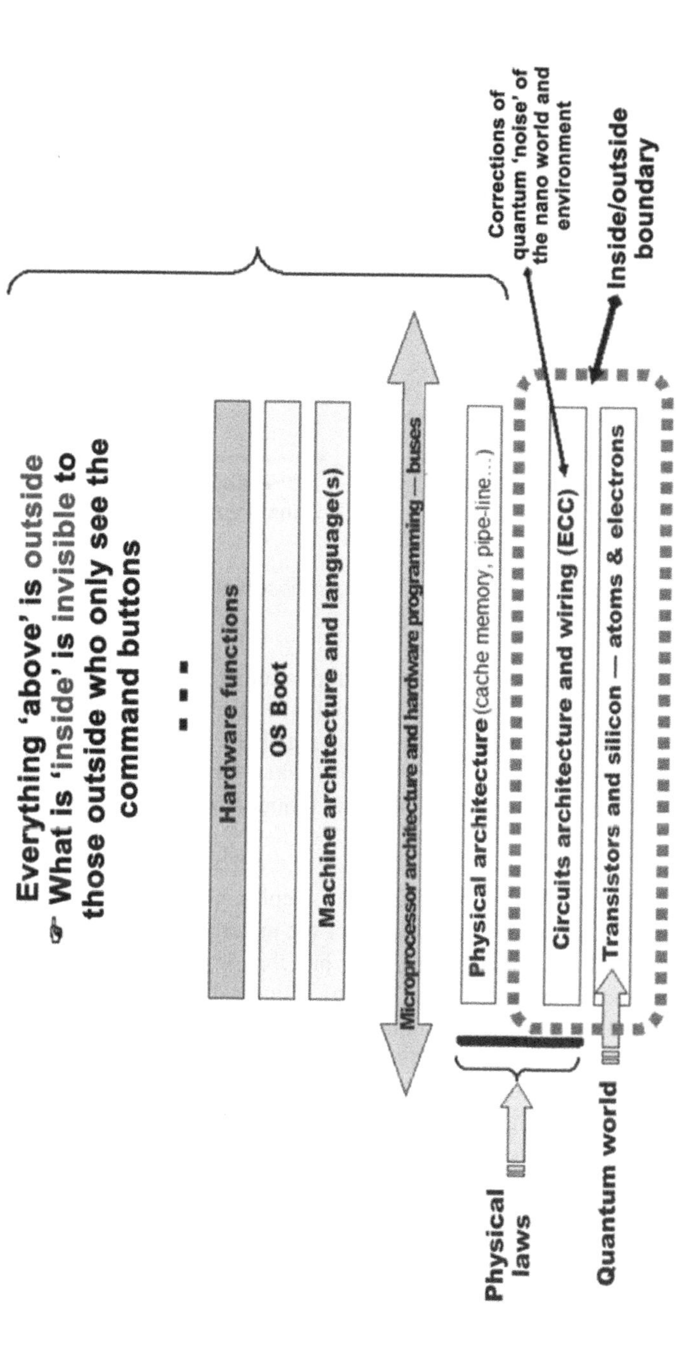

Figure 3.7. *Organizing the inside/outside boundary*

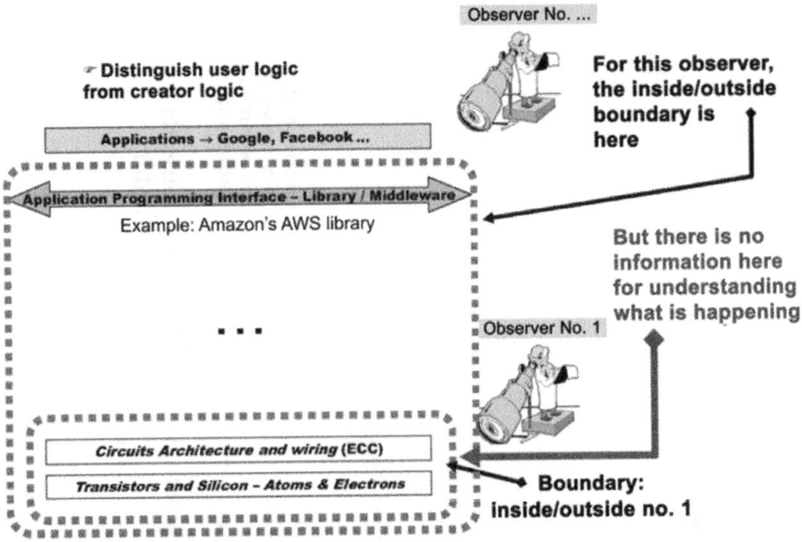

Figure 3.8. *Confusion between observation levels*

If the most external observer receives information from the intermediate levels that they may not even know about, they will be unable to understand what is happening, and vice versa[9]. An observer who gathers events and/or information about the internal levels, triggered by actions taken by external users, would be completely unable to understand what the user (about whom they know nothing, by definition) is doing.

As physics has taught us since Copernic, Kepler, Galileo, Newton and some others, any measure taken requires a theory of the underlying phenomenon: without the theory, we are blind. It was when Kepler had the idea that observed trajectories could be something other than circles, namely, ellipses, that observation data about the planet Mars obtained by Tycho Brahe suddenly seemed to light up from within.

Programming, in the widest sense of the term, offers a multitude of examples of what an under-trained programmer risks. In the transactional notation, the first property is the perfect market, the A of ACID[10]. This means that, to understand what is happening in a system, we must above all not look at the details of operations generated by a transaction because this detail is circumstantial, depending on the

9 Certain characteristic errors of networks such as the *time out*, meaning that a time constant was exceeded, have no meaning outside the network sphere.

10 See: https://en.wikipedia.org/wiki/ACID: atomicity, consistency, isolation, durability.

available resources and the level of system saturation (Figure 3.9). The execution of the basic processes of the transaction is generally non-deterministic, which does not affect the overall result of the transaction[11]. This same kind of mechanism also exists in operations used to interact with a network. The circulation of packets follows the availability of network resources; the only thing that counts is the fact that nothing is lost, whatever the path taken, and that the overall operation follows the accepted time limit.

In the case of a transaction, the only demand a programmer has is to know if the action, the result of a transaction, was completed or not. The details of the operation are unimportant and, in this case, the task of programming is indeed simplified through the generalization of an abstract construct that hides useless details from a programmer of a certain level, which comes down to distinguishing the static side of the program, in other words its organization, from its dynamic side, in other words its behavior.

Seen through the lens of this new abstraction, the execution details are background noise, which would reduce the programmer's efficiency if they were shown to them. It is the role of the transaction monitor to hide the combination from the programmer, despite a few rules that need to be followed, without which the operation would be unprogrammable because it would be too complex; this is the I of ACID. In Figure 3.9, if the transaction requires the execution of 1,000 functional instructions, the number of possible interruptions is 1,000!, that is, a colossal number, on the order of $5 \times 10^{2,567}$, and entirely inaccessible. The interruptions can be prepositioned explicitly by the programmer using ad hoc synchronization commands – see section 3.3.3 – or left to appear to the transaction monitor, which will decide depending on the state of resources. The effect of interruptions is to make execution delays variable, from a minimum of D_{min} to an acceptable maximum beyond which the operating system will detect an error; hence, resources considered critical for performance must be monitored: to improve execution flow. This is a new variation of Laughlin's "deceitful turkey". The noise is however functional for the layer below, which controls the sequence of detailed operations; in Figure 3.5, this corresponds to the layers called middleware.

To conclude: on the basis of these few examples, we can understand why, in the absence of models, the diagnostic of observed anomalies is impossible because we do not know where, nor what, nor how to search; we are faced with a puzzle whose final image is unknown, and with irregularly shaped pieces to boot! Now, we only find what we seek when we know how to seek. The model acts like a filter for the real, which is not given but constructed by the designer according to the internal and

---

11 For details about these fundamental mechanisms, see Chapter 9 of *Architecture logicielle*, 3rd Edition, Dunod; or Chapter 8 of *Science of the Artificial*, by H. Simon.

external FURPSE/PESTEL constraints, and even the economic project constraints, CQFD (Cost, Quality, Functionality, Development times) and TCO (Total Cost of Ownership, including maintenance). This approach, illustrated here by the computer stack, is entirely general.

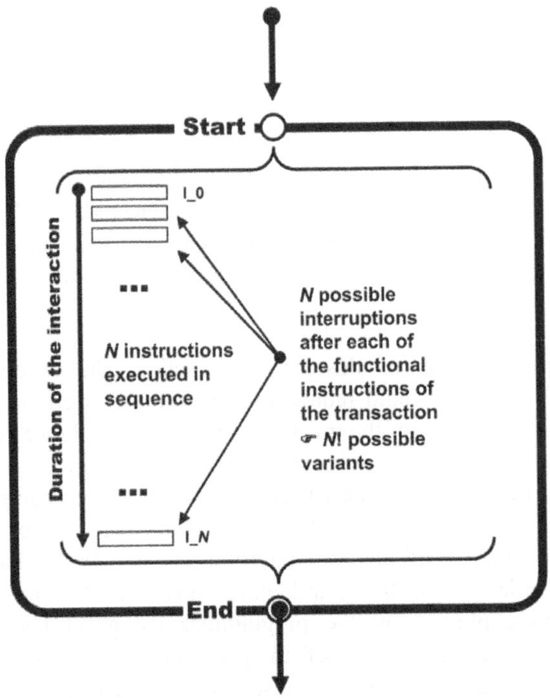

Figure 3.9. *The middleware layer of ACID transactions*

REMARK 3.3.– Without the theoretical BEH boson model, physicists working at CERN would never have been able to find a trace of the Higgs boson whose particularly short lifetime is approximately 10^{-22} seconds, among the hundreds of millions of events detected by the LHC measuring instruments. Even with the help of the most powerful computers, this would have been useless given the combinatorial nature.

For the same reasons, we can, in the absence of models for brain function at different levels, doubt the putative results of massive projects such as the *Human Brain Project*, since we end up faced with 100 billion neurons (10^{11}) and more than a million billion connections (10^{15}) for which we do not know the functional/organic inside/outside boundaries, nor how the cerebral stack is organized, even if we have a vague idea of functional zones that, compared with the computer stack, correspond to the level for applications. It is obvious that the brain possesses powerful error correction codes and many functional redundancies that are certainly not distributed randomly given the reliability of this organ that fascinated von Neumann so much. But where are they, physically, and how do they work? For the time being, this remains a mystery[12].

Von Neumann recommended that corners must not be cut[13]. This is what seems to have been forgotten, so beware Laughlin's "deceitful turkey" and Wittgenstein's aphorism, number 3.328, for MRIs naturally produce a profusion of data whose meaning remains hypothetical in the best of scenarios, and most often, totally meaningless in the absence of a model; just as with the canals on Mars, "discovered" by the Flammarion siblings and which turned out to be mere artifacts of the instrument used at that time. The intensive use of computers does not at all protect us from this kind of trap. In his rather frank way, Réné Thom would say that, "our problem is the insignificance of the data we collect" – a word of caution that is too often forgotten.

3.3. Milestones from the history of computer stack development

The history of calculating with the help of computers is sufficiently recent and concentrated in both time and space. It is essentially around MIT and within Route 128, which encircles the city of Boston, that we can provide an overview to help us understand what is at stake[14]. It really begins with the widespread use of analogue computers throughout World War II (B29 target viewfinders were equipped with an analogue calculator) and continues well after, since the calculators in the supersonic Concord were also analogue machines. For whomever wishes to understand in depth how this fantastic technology developed, it is particularly instructive to look at this history which is barely 70 years old, or the equivalent of two to three generations of engineers, with the progressive construction of the computer stack as the thread that connects everything.

In an analogue computer, there is no computer interface stack. The user is faced with electronic circuitry whose physical function they must know intimately in order to be able to use it with without making errors; it is the same for computers such as

12 See the recent books by Dehaene, *Les neurones de la lecture* and *Le code de la conscience*.

13 In the introduction to Chapter 1 of his book *Theory of Games and Economic Behavior* (Princeton University Press), a theme which he revisited in his conference at Columbia University in October 1954, reproduced in Volume VI of the *Complete Works*, *Method in the Physical Sciences*.

14 For more details, see the book by Goldstine (a colleague of von Neumann at Princeton), *The Computer, from Pascal to von Neumann* (Princeton University Press).

the ENIAC and the EDVAC. Nothing about them resembles programming in the slightest; the user is a physicist/engineer who must know how to interpret the measures. This approach spread until the 1960s with electromechanical materials using perforated cards, like those developed by IBM and Bull.

With the first computers using the von Neumann architecture, like the Whirlwind machine from the SAGE project based on the ECP prototype developed at the IAS in Princeton, things changed radically. One of the virtues of this architecture is to disconnect the electronic circuitry that provides transduction from the commands that initiate the transduction. The existence of memory makes it possible to enter commands as a series into the memory, including the data to process; this is how the first saved program was born. In this context, programming means: 1) to functionally know the available commands, or the *what*, independent of their electronic manifestation, or *how* it is done, 2) arranging the commands in an order to create the required function and then 3) depending on the result (which is a feedback), selecting the next command to be completed.

This is how the first flowcharts appeared, trials of the first programs, which were certainly written by von Neumann himself. This approach made concrete the fundamental distinction between external language/internal language underlying this young technology that von Neumann described in his final book, *The Computer and the Brain*, just before dying.

The equally fundamental fact that commands (the instructions of the computer, called *statements*) and data share the same memory made the birth of the first programming languages possible. These were called "assembly languages" for obvious reasons. The programmer, instead of seeing only 0s and 1s, could use mnemotechnic codes built with the much easier to use alphanumeric alphabet (26 letters, punctuation, numbers, and special characters such as =, - and ×; at the time, it was a serious difficulty given the rarity of resources and the cost of electronic equipment). The so-called assembly programs, ancestors of the compilers for more evolved languages, converted the alphanumeric codes into a sequence of 0s and 1s. Doing this, programmers could write their first programs using symbolic codes that the program would replace by the binary codes of the computer. All things otherwise being equal, the hardware architect could modify their coding system, depending on the hardware constraints, independently of what the programmers wrote. This is how the first translations were born. The programmer's logic thus became disconnected from that of the hardware designer; they could thus work separately, with programming languages now functioning as the mediator of this fundamental interaction that has since structured its history (Figure 3.10). Defined as such, the interface carries the logic of integration that makes it so a level L can "forget" how level L-1 is made, which could be written symbolically as

$[1 \otimes 1]_{Level(L-1)} \xrightarrow{Integrate} [1]_{Level(L)}$, meaning that the complexity does not multiply, which is a fundamental property of the interface.

REMARK 3.4.– This is also something the conscious brain does with ease when it interacts with other brains via language (see Figure 2.3).

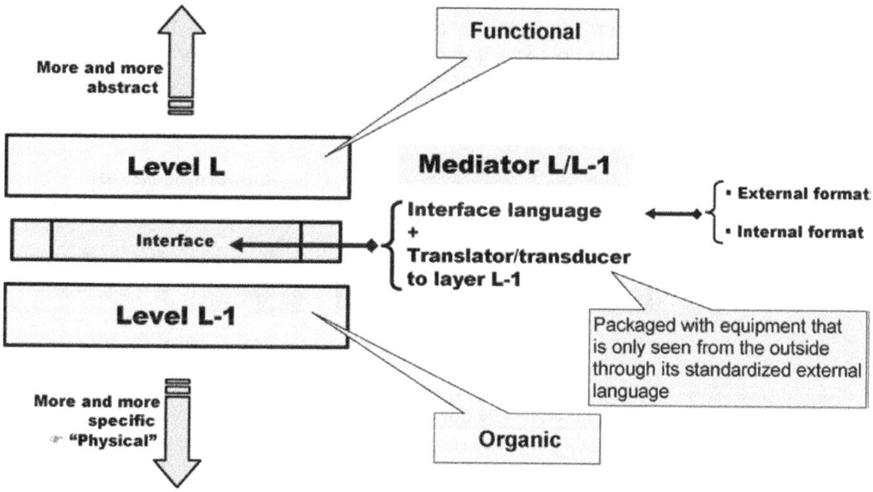

Figure 3.10. *The functional/organic mediator*

3.3.1. *Digression on interface engineering costs*

Let us give some factors in the engineering costs of developing this interface to fully illustrate the breadth of the conceptual revolution brought about by the logic of the stack by looking at an interface that is known by all nowadays: the hardware/software interface and the languages associated with this layer.

At the level of the hardware/organic layer, the computer is a world of 0s and 1s that serves to code information, data and/or instructions, in the binary language associated with the electronic transducers at this layer, and this has been the case since the beginning; it was a choice made by von Neumann himself for the Princeton computer at the IAS.

There are always two ways of using this fundamental interface of the computer, as with all the interfaces of the stack, for that matter (see Figure 3.11):

– the "assembly" approach, which is a coding system contained *within* the layer, analogous to stenography;

– the evolved language approach, which is a completely abstract reformulation of the functional capacities of the organic layer, independent of the transducers that constitute it, for *outside* users, via the programming interface.

In the first case, the programmer creates texts in the internal language (IL) of the layer. The mode of expression is thus dependent on the physical organic structure of the layer. In the second case, the programmer creates a text that is a reformulation of the problem in the most appropriate external language (EL) for resolving the problem in that layer (FORTRAN, COBOL, C/C++, Java, C#, PYTHON, or even a specialized language adapted to the needs of the users, such as SQL for data management).

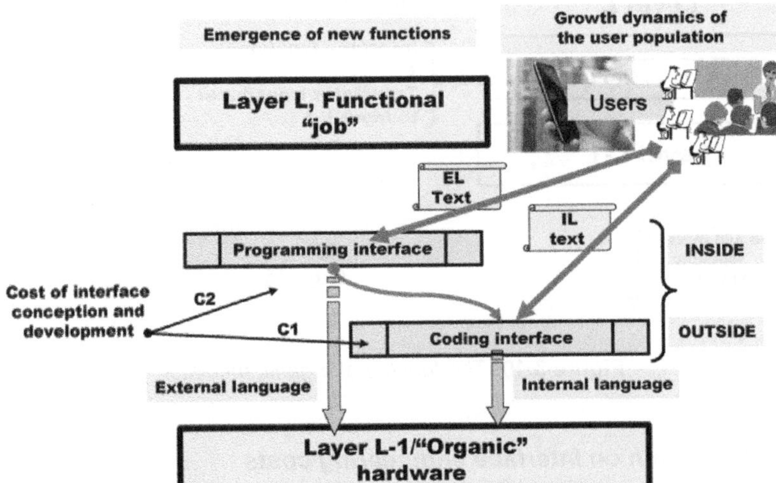

Figure 3.11. *Inside/outside interface costs*

Niklaus Wirth, the inventor of the languages PASCAL and MODULA, summarized the activity of making programs, that is, what is usually called "programming," with the following formula[15].

Programs = data + algorithms

This famous formulation underscores the dual activity in programming (working with data, on the one hand, *and* working with algorithms, on the other hand, both being inseparable and in a symbiotic relationship), in other words, the transformations that will be performed on the data in such a way that builds, step by

15 See: https://en.wikipedia.org/wiki/Niklaus_Wirth.

step, instruction by instruction, the desired result, what Wirth qualified as successive refinements[16].

In the first case, a programmer makes a binary image of the computer's memory, word by word, 32 or 64 bits depending on the architects, with only the help of mnemonic symbolic codes that are well known by programmers with assembly experience, but with the major inconvenience that the text obtained in this way depends entirely on the structure of the computer, without any possibility of generalization or evolution[17]. In the second case, the programmer can temporarily "forget" about the structure of the computer to concentrate on an "elegant" formulation of the problem at hand given the more or less foreseeable evolutions in the environment, including that of changing computers, and interactions with colleagues. It is up to the compiler to create the appropriate construction of the binary image.

It is interesting to analyze the consequences of these two approaches on the potential population of users accessing the organic layer. The dynamic of this growth, seen in Figure 3.12, corresponds exactly to the classic model by Lotka–Volterra, of which it is an excellent illustration[18].

With successive models of computers coming one after another at the rate of Moore's law (see below), the durability of the texts in IL is necessarily limited. To keep up with evolutions in technology, programs must be regularly rewritten at the assembly level. Moreover, the skills acquired by a programmer also expire at the same rate; they have no economic value, hence the hereditary term that quickly appeared. IBM got around this difficulty by progressively standardizing the hardware/software interface in its computer range models called 360[19], but that nevertheless blocked programmers from migrating towards other types of materials.

The rapid appearance of evolved External Languages (EL) such as FORTRAN and COBOL made the growth in the number of programmers in IL even more

16 This is the title of one of his books, *Algorithms + Data Structures = Programs*, published by Prentice Hall; it is easily downloadable.

17 Today, this corresponds to less than 1% of the population of programmers, estimated to be around 26 million in 2022.

18 See: https://en.wikipedia.org/wiki/Lotka%E2%80%93Volterra_equations on the Lotka–Volterra equations, https://en.wikipedia.org/wiki/Alfred_J._Lotka on Alfred Lotka, and Lotka's book *Elements of Mathematical Biology* (Dover Publications, New York, 1956. Originally published as "*Elements of Physical Biology*" in 1924).

19 An extraordinary performance by the engineering teams at IBM that must be recognized as such, but it was obviously not enough. See: https://en.wikipedia.org/wiki/IBM_System/360 and the history books at MIT Press in the series *History of Computing*.

problematic. The use of EL completely resolves the problem of expiration as long as the compilers that translate EL to IL follow the cadence imposed by Moore's law. Texts written in EL became durable, along with the skills acquired by programmers; everything became cumulative, and programmers quickly began to prefer using evolved languages, which were considered much more economical.

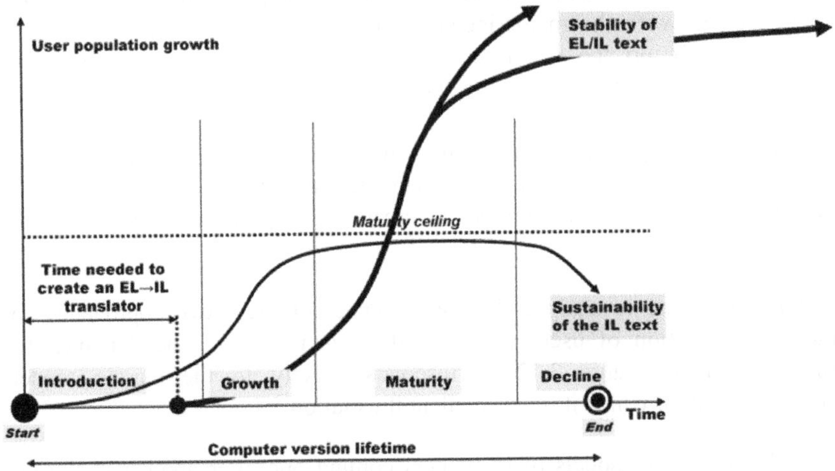

Figure 3.12. *Growth rates in the population of programmers*

The fundamental parameter of growth in the population of programmers therefore came from a balance between costs/development times/qualities of the translator from the EL to the LI, versus the capacity and interest of programmers in using a particular EL rather than another, with the evolved languages appearing simpler at first. Hence, the Ada language never managed to take off, since besides the difficulties in adapting Ada compilers and in learning the language, programmers always preferred the pair C/C++, which were considered simpler, as well as "object"-oriented languages such as Java, C# or PYTHON.

A few words to conclude this digression on the costs of engineering these different translators. The costs of development for an assembly IL have never surpassed 10 ha – even for computers containing games with complex rules such as IBM or Bull computers, or even Intel microprocessors – and often cost much less[20]. The production of an assembly program does not require particular knowledge beyond the complete mastery of the hardware/software interface; on computers such

20 Habitual counting measure in software engineering, corresponding to a programming engineer working for one year, or approximately 220 days, or 1,607 hours in France (after the law limiting the work week to 35 hours). 10 ha, rounded, makes for about 20,000 hours of engineering work.

as IBM or Bull, this corresponds to a documentation of more than 2,000 pages, for games with rules that could include several hundred different operations. The issue of external languages considered "high-level" EL or third generation is an entirely different issue. For beyond an intimate knowledge of the computer, including the caches, the instruction pipelines and the data, we must master the theory of languages and compilation techniques[21]. This is the equivalent of another two or three extra years of reasonable abstract study after a master's degree, followed by an equally long amount of professional experience that can only be acquired by making compilers, not by reading books[22].

For languages like FORTRAN or COBOL, teams of at least 10–15 engineers are needed, for three or four years, resulting in an initial cost of about 40 to 50 ha, for a version that would then need to evolve given Moore's law and the need for standardization. For a complex language like Ada, it would be closer to 70 or 80 ha. And since several different EL are needed to fulfill the different programming needs of users and professional clients, an overall team of up to 50 or 60 engineers would be necessary for the languages that the computer manufacturer wants to be able to employ. More if we integrate Data Base Management Systems (DBMS) teams, which would also be strongly influenced by compilation techniques in which languages like SQL play a fundamental role of the same kind as for users of databases. This is also the case since the 2000s, after the arrival of the so-called "multicore" microprocessor designs, currently being used more and more, whose needs are varied and plentiful, in everything from smartphones and video games to the most advanced weapons systems such as combat drones.

The principal lesson to take from the invention of evolved languages is that without them, ICT would never have had the exponential development that we have witnessed. In the 1980s and 1990s, there was still a good dozen computer builders and editors, each with one or more teams as the ones described above, and we could say that several hundred engineers (in simpler terms, 20 teams of 60 engineers only gives us 1,200 developers working on compilers!) made it possible for millions of programmers to work at ease by liberating them of the obligation of knowing the idiosyncrasies of different common computers inside and out. Without them, the Cloud designers and the multicore microprocessors that we have today would be

21 They are no longer taught in France or even Europe since the 1990s; we may as well say that the know-how has been lost!

22 For a more detailed overview, in French, see the second part of *Architecture logicielle*, or the authors' website; reference works are exclusively in English.

completely unprogrammable. The leverage went from 1 to 1,000, or even 10,000, which is, literally, absolutely stupefying and had never been seen in history of technology.

The same lesson applies to the other interfaces, namely the stack for accessing networks, the stack for accessing external memory via DBMS or the graphic user interface (GUI) that "Thumbelina" makes use of without realizing how much technology is behind these "miraculous" interfaces. But without these languages, without the theory of languages and without compilation techniques, there would be no Information Society.

3.3.2. *Observing the nano-world*

We can remark, in passing, the possibilities for observation that specific interfaces offer, as in Figures 3.10 and 3.11. The existence of an EL makes it possible to define a level of observation that can then be translated into the organic structures of the L-1 layer, even if this layer is composed of nanometric functions that are unobservable to the naked eye, and even if it was magnified with a light or electronic microscope. The instrumentation apparatuses that are added in this way are observation probes at the IL level, from which we then only have to gather the measures so that by inverse translation, we can translate what was gathered into the EL that only the observer is capable of understanding. Everything then becomes a matter of correspondence between languages and reversible translations/ transductions which, if the languages are well made, are mathematically verifiable. Thanks to the theory of languages and to Turing's genius intuition now materialized in the Church-Turing theory, we have the possibility of working in the nano-world through the process of perfectly controlled, interposed translations, and with the same level of security as in the world at our scale[23].

All of this falls apart if observation is not done at the interface itself, for then the flow at the *inside/outside* boundary interferes with the observation of signs since it permits the entry signs of unknown origin and therefore stripped of meaning for the layer receiving them. This is more evidence of the crucial importance of this boundary, of the self and the non-self, as specialists of the immune system would say[24]. It is naturally the case at all the levels of the computer stack, at all levels of abstraction, therefore illustrating the need for an autonomic approach and autonomic

23 The formalism of Lambda-calculus, Church, and that of Turing's machine, although very different, are nevertheless equivalent; see the distinction between functional languages versus imperative languages.

24 See the book by our colleague at the Collège de France, Philippe Kourilsky, *Le jeu du hasard et de la nécessité* (Kourilsky 2014) and in particular, the Conclusion of this book.

components. We know from the work of Bertrand Russell on logic and types that the confusion of levels mechanically creates syntactical paradoxes that make for inconsistent reasoning, and that we therefore must avoid in the construction.

3.3.2.1. *Acceleration of history*

From this point, the history speeds up, as much on the side of hardware as that of software, depending on technological innovations such as the discovery of transistors, which would become the dominant structure of hardware towards the end of the 1960s and beginning of the 1970s; they progressively eliminated all others until we ended up with entirely *transistorified* computers such as the PDP-11, in 16 bits, from the Digital Equipment Company (DEC), which now no longer exists[25].

In the meantime, the structure of computers also saw significant evolutions in terms of time management. At first, everything was synchronous, and there was only one clock, which meant the computer would mechanically work at the speed of the slowest element. This is the case for the parts that let the outside world interact with the computer, such as perforated card readers and printers, which are what we call entry/exit peripherals, and even, at the very beginning, symbionts, by analogy with what we see in certain living organisms. The same piece of hardware could thus house several processors, each working at the rhythm of the equipment they were managing; it was up to the programmer to synchronize the operations when needed using ad hoc commands. A computer such as the Gamma 60 by Bull was one of the first to use these mechanisms systematically[26].

Computer assembly languages became varied and were complemented by commands that were specific to synchronization, such as the semaphore languages used by sailors and harbor masters to manage the entries and exits of ships in ports. One of the main texts of this period was the report "Cooperating Sequential Processes" by E. Dijkstra, written in 1965, and which can be easily downloaded from the Internet. The programs written by programmers are compared to sequential processes that synchronize with each other thanks to the capacity that would come to be known as Inter Process Communications, the ability to communicate and synchronize among process image programs; in the jargon of object-oriented languages such as Java or C#, these are the libraries of RMI (Remote Method Invocation) functions, still not coherent among themselves.

This new way of breaking up operations led, from the 1960s onwards, to an entirely new industry that took on the development of these capacities, and gave rise

25 See: https://en.wikipedia.org/wiki/PDP-11.

26 See Bourboulon, J. (2005). *L'aventure gamma 60*. Hermès-Lavoisier, and the website: http://www.feb-patrimoine.com/, which contains a lot of historical information.

to operating systems, themselves structured in layers, without which programming would have remained an inaccessible exercise and a practice limited at most to specialists, making the developments that came later problematic.

One of the flagship systems of this time was the MULTICS system developed at MIT and commercialized by Honeywell-Bull in the 1970s[27]. It was with MULTICS that the computer stack became stable in the form it still has today. In particular, the "recursion" level that can be seen in Figure 3.3 became standard and independent of the languages translating the concept of recursiveness, with, in certain computers like the DPS7 model by Honeywell-Bull or the Intel 432 microprocessor, an ad hoc instruction set for managing the levels of the corresponding stack. Previously, the recursive stack was seen as a priority in a programming language, and therefore specific to any one language, which meant that several languages could not be used in one application. Separating it from the language by putting it at the level of the computer made it into a generic property, which would simplify the work of programmers; once more, the principle of "generalizing to simplify" worked. Parallel to this evolution, operating systems were supplemented with libraries of ever richer and varied functions, of which some were directly visible in programming languages, like the scientific libraries associated with the language FORTRAN (hence the general need for a recursive stack)[28].

From the 1980s, all of the ingredients of the computer stack in its current version were already in place and became the basis for the last great computing mutation, that of distributed systems and client/server architectures (see the "Integration Middleware" level in Figure 3.5). The mostly parallel architectures combining thousands of active processors either in the form of a Cloud or in the form of multicore processors is nothing but an extension of what was invented in those years, with an added change of scale.

In all of this evolution, the computer stack as it is presented here in its simplified form played a major role and served as a model for the designers of telecommunications, databases, human–machine interfaces and the great technological infrastructures such as the network for transporting energy or communication infrastructures. It is the integrative structure par excellence that allows different occupations to be coordinated and have their actions converge, a structure without which disorder would be inevitable.

---

27 See Organick, E., *The Multics System: An Examination of Its Structure*, available at: http://mirrors.acm.jhu.edu/textfiles/bitsaved/Books/Organick_TheMulticsSystem.pdf.

28 See the second part of *Architecture logicielle* and the authors' website for more information on this entire sequence.

3.4. Moore's "law": a structure for integration

Among the first computers and those that we use today, the increase in performance has been so spectacular (an improvement of a factor of 1 million) that it remains without precedent in the annals of the history of technology. One of the very first computers built for the SAGE system under the supervision of Jay Forrester at MIT, worked at the rate of 20,000 operations per second; the average time it took for it to work well was about 20 minutes[29]. Whirlwind used 18,000 electronic tubes as active components, each consuming 50 watts, the equivalent of an electric lightbulb of medium power, or 900 kW wasted! Since then, transistors have replaced electronic tubes and there are billions of them, but they still consume energy.

In 30 to 40 years, things went from extreme shortage and problematic reliability to abundance and the quality we now know. What appears from all sides like a miracle is in fact the result of exemplary engineering that owes everything to the intelligence of its creators.

This growth in performance was theorized in the form of a "law," known as Moore's law, by Gordon Moore, the founder of Intel[30]. It is nothing like the laws we know from physics. It is a simple statement resulting from the observation of performance, which doubled every 18 to 24 months in terms of the precision in engraving, which led to greater integration, facilitating the work of designers. It has proven to be rather accurate up until now (see Figure 3.13).

This notwithstanding, the law itself does not explain anything; it is a simple correlation, which we know to be misleading on occasion[31]. A correlation does not mean causality, something which some forget when it is convenient.

Equipped with the constructive logic used in developing the capabilities of the computer stack, we get a much better idea of the *how* of this unequaled transformation in which, starting from a point of extreme rarity, we arrived at a situation of abundance with the spectacular results we are all familiar with, but nonetheless with some serious reservations when it comes to matters of security.

---

29 See: https://en.wikipedia.org/wiki/Whirlwind_I.

30 See the detailed article by Ethan Mollick: Mollick, E. (2006). Establishing Moore's Law. *IEEE Annals of the History of Computing*, 28(3), 62–75; as well as: https://en.wikipedia.org/wiki/Moore's_law.

31 See: https://en.wikipedia.org/wiki/Spurious_correlation and http://www.tylervigen.com/spurious-correlations, a website that lists some of the most misleading and spectacular examples.

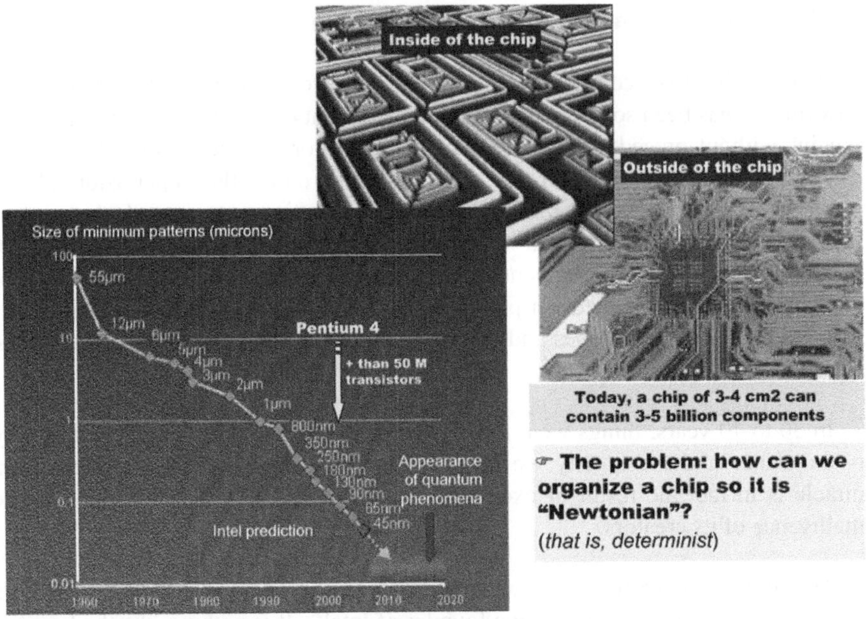

Figure 3.13. *Moore's law and engineer performance*

A mismanaged abundance can quickly become synonymous with waste, a well-known phenomenon in economy, where we sometimes speak of a resource curse in states that benefit from them and who, because of these resources, have no motivation to work seriously. Today, we would surely have to dedicate a part of the available computing resources to a serious study of security, as was done at the time of MULTICS by MIT and the MITRE Corporation, a time when it was precisely the lack of resources that prevented the realization of what they had imagined. The VAX computers by DEC, with their VMS system, had benefitted from it, but DEC has now disappeared.

The logic of the stack had the principle virtue of channeling efforts, on the one hand by augmenting the number of levels, which led to an increase in vertical abstraction, and on the other hand by stretching out the specific capabilities of each one by diversifying them horizontally in order to organize the needs of users/customers more efficiently and find a better place for proposed functions, all with the ultimate goal of optimizing the overall performance of the ensemble.

We will now provide some examples of the execution of this logic illustrating the central organizing role of the stack.

The continuous progress in the physics of the components made it possible to spectacularly increase the number of transistors per cm^2 – this is what the creators of circuits called integration, large-scale integration (LSI) at first, then very large-scale integration (VLSI) when the number of integrated transistors reached the hundreds of thousands, then millions. It was a time when one central computer unit, the equivalent of 200,000 to 300,000 transistors/components, compared with the 18,000 of the Whirlwind computer from the 1950s, could fit on just one circuit. The abundance of transistors certainly allowed an increase in memory capacity, but it especially made it possible for memory to become more reliable by replacing the simple parity checks on 1 bit that detected errors using the error correction codes based on the theoretical studies by Shannon, codes that not only detected the error but corrected them automatically. For this, a capacity of 7 to 11 bits per 32-bit word must be reserved, or approximately a 30% redundancy, for central RAM, and much more for disks that are more exposed to transmission errors. Consequently, the availability of computers, which only lasted a few hours in the 1970s (the reason for the resuming point mechanisms managed by operating systems and the restart mechanisms "from cold" or "while warm"), went to several years in the 1980s, almost "erasing" the omnipresence of errors, perhaps giving the false impression that failures had disappeared even though they were still there! This is an intelligent use of the principle that "more is different" and of the implementation of the third principle of systemics concerning the compensation for errors.

REMARK 3.5.– We should note that in the formula that gives availability $D = \frac{MTTF}{MTTF + MTTR}$, if $MTTR$ is equal to 0, because of autocorrection, availability becomes $D = 1$; this is the same approach that is used in all the mechanisms for "failure tolerance."

More generally, the abundance of transistors made it possible to implement the functional redundancies that von Neumann had studied theoretically. What stands out in this history is that the theoretical models were often ready before the technology was, which is a rather unusual situation in the history of technology, where it is often the inverse, as for example with steam engines, which were invented a century before thermodynamics. We can say the same thing about programming languages, which were largely anticipated by the work of logicians in the 1920s, such as Russell, Whitehead, Wittgenstein, Gödel, Herbrand, Turing, Church, Kleene, Gentzen and Kripke.

In Figure 3.3, at the level called "physical architecture," we see specific functions called "memory cache," "pipeline," etc., all of which are acceleration mechanisms. The memory cache is the simple ability to memorize, making it possible to avoid having to systematically recopy the information to the central memory, which is much slower (by a factor of 1,000 or more) than the one available in the chip. By doing this, we violate one of the fundamental principles of data management, which is to never duplicate information. This means that managing this cache, as with any other cache, requires special precautions and a great amount of experience with the corresponding algorithms[32]. It is typically a function that must be centralized, entrusting its construction to the most competent systems programmers so that in return, the entire community using the stack can benefit.

If there is an abundance of transistors, the architect of the computer can reserve a portion of the transistors to use as caches of data and/or caches of instructions, so that systematic recopying, which is the only thing that guarantees coherent processing, can be avoided. With chips containing one billion transistors or more, most of the chip is in fact memory. The cache thus plays the role of cached registers, which means that the notion of registers itself no longer has much use when there is no limitation in size other than what is permitted by technology. By doing this, optimization algorithms for the use of registers in the computer, a nightmare for compiler architects, becomes a non-problem! And the overall complexity can be reduced, an organizational "miracle."

More interesting than this is the ability called a "pipeline," which has existed since the first computer imagined by von Neumann himself. First, it was the simple ability to reuse hardware based on the fact that decoding instructions was a process with stages. While one instruction is being executed using the ad hoc transducers, we can begin to upload the next one and put away the previous one; this means that the same hardware can be used in different contexts, hence, incidentally, the difficulty in understanding the pathways used if we do not understand how "it is done". The structure of the instruction set, a fundamental interface at the hardware/software boundary, is what determines the size of the pipeline. This interface is visible to systems programmers, which means the discovery of a solution that is a balanced compromise requires a dialogue between compiler architects, representing the community of programmers, hardware architects and technologists, as was the case in the 1980s with RISC architectures. The consensus is built around the stack, which makes it possible to ascertain the overall balance of the situation. This kind of pipeline creates a synchronous parallelism, made possible by the ad hoc organization of the way the instruction set is coded.

More subtle than this is the asynchronous parallelism, the "real" parallelism, that appeared in computers towards the end of the 1980s with VLSI. The abundance of

32 See the three volumes of *Art of Computer Programming* by D. Knuth.

the transistor resource was such that we could not only increase the size of cached memory, but also multiply the number of functional units, that is, transducers, which were necessary for decoding and for the execution of instructions[33]. This is how pipelines with 15, 20 and up to 30 levels came about, as in Intel's Pentium IV. Instructions arrive in the pipeline as long as they are not in conflict with each other, and this continues until the pipeline is saturated. Without anything being changed in the code, users see the performance of their programs improve with factors of × 2, × 3 until other phenomena come to break the sequences induced by conditional `IF` *cond* `THEN` ... `ELSE` ... instructions, which stop the pipeline from working. Intel went as far as inventing so-called speculative pipelines where the processor tries to execute two branches of the alternative until the result of the condition validates the branch that will be needed to continue processing. By doing this, the complexity of the hardware was increased beyond what is reasonable, which brings about new problems that are even more difficult to resolve, thus violating Turing's own advice: "Keep the hardware as simple as possible." Here again, we see how the stack facilitated the emergence of a dialogue between specialists, making the discovery of balanced solutions among the interested parties possible.

It is a waste of energy, and a risk in feasibility, to try to correct what often turns out to be bad programming using these kinds of mechanisms. It is better in this case to train and educate users of technology to realize healthy program architectures, that is, testable and maintainable ones. The stack thus became the center of a virtuous cycle in which everyone sought to improve themselves for the benefit of all. This was a true, mutually beneficial relationship, following the logic of the prisoner's dilemma, well known by strategists in game theory. It is the most intelligent way to organize complexity.

In the 1980s and 1990s, hardware engineers had to face a double rupture: 1) that of manufacturing times for VLSI, which lasted several months, and 2) that of exhaustive verification of certain material aspects of the circuit, such as the wiring and the junction points of the connectors, equaling hundreds of kilometers worth of wires buried in the silicon substrate (see Figure 3.13).

With VLSI/SOC and the manufacturing time of circuits, the cost of a hardware bug became prohibitive. The manufacturing process was organized so that at the end of the production line, all selected circuits on the wafer were satisfactory from the materials and the electrical points of view, but if the functional logic, that is, the circuit after being integrated, did not do what it should do, it was necessary to modify it and restart the entire manufacturing cycle since the circuit could no longer be corrected as was done in the previous generations using very visible patches.

33 For example, in the designs called VLIW; see: https://en.wikipedia.org/wiki/Very_long_instruction_word.

Statistically, a design error would cost 0.5 times the duration of the manufacturing cycle, to be multiplied by the number of people put on hold, to which we must add the loss in profits due to the delays. In industrial projects, this quickly leads to hundreds of people at a standstill, and proportional losses in revenue; it is disqualifying.

The moral of the story is that functional validation must be anticipated before even beginning the manufacturing cycle, and for that it is necessary to use simulation techniques with ad hoc computers from the preceding generation[34]. This is a classic model in the history of technology, where computers and systems from generation N make it possible to improve computers and systems for generation $N + 1$ so that they perform better. In the field of circuitry, hardware description languages were developed, for example, VHDL, that made it possible to do validation work on the blueprint text using all acquired technical skills, which reached perfection in the 1980s. Once the validation is acquired, there is nothing left except to compile the validated text through what was named "silicon compilers," which enabled the production of everything needed for the manufacturing process, including the tests and diagnostics of the intermediary manufacturing steps. This is how we came to see the real power of the concept of an external language/internal language with the transducer as the mediator for interaction, providing a means for mechanically manufacturing the internal form from an external abstract description (see Figure 3.10).

In chips such as those of the Intel Pentium IV generation, which contain a total of over 50 million transistors, it would no longer be possible to "manually" verify that everything is correct, even with the means for observation such as electron microscopes. There are too many connections, and the engraving is too fine. But the description language lets us generate exhaustive tests that can be executed at the speed of computers of the previous generation. The designers, by virtue of this external language, thus replaced mechanical "manual" verification with an automated computer verification that occurs at the speed of the computer. In this approach, the engineering process itself became mechanized, along with the tests and diagnostics validating the results of the process. Everything is thus dependent on the quality of the conception models and tools, which are defined and set to work by systems architects. All these tools require a great familiarity and serious experience with the overall process, which cannot only be learned in books, but primarily through experience and mentoring, over time and through successive generations of computers.

34 See the books by Maurice Dumas and Bertrand Gille, already cited.

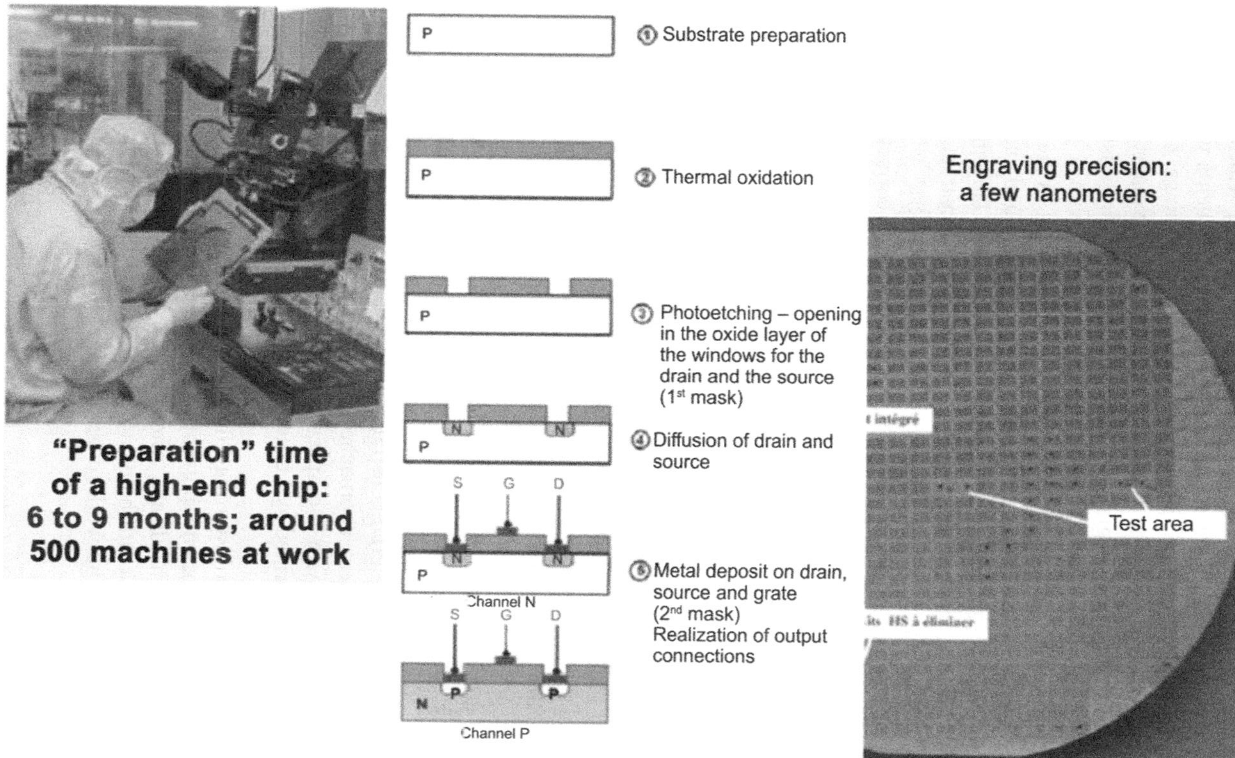

Figure 3.14. *VLSI engraving*

The diagram in Figure 3.14 gives an idea of the engraving techniques that make it possible to connect transistors to each other in order to create the logical functions needed for operations carried out by the computer. Starting with the Pentium IV generation, engraving was done over a number of layers in such a way that made it impossible to analyze defects immediately afterwards. Things had to be done correctly from the start, ensuring validation with the help of exhaustive tests created according to what had been engraved. Since these are robots in finished states, there are naturally numerous cases to verify, which is doable with a sufficiently fast computer, but would be totally impossible "manually" (as is the case with the reverse engineering of the complete circuit), since to observe and understand what is happening, it would require the destruction of the nanometric labyrinth of the wiring framework. This rigorous engineering is the apex of constructive methods of engineering, sometimes called co-design or design to test, which show that complexity can be organized and controlled when it is done correctly and the tools and methods that correspond to the required quality are used.

In this kind of approach, we note that complexity such as we have defined it (complication, complexity, incertitude, ignorance) is organized in the following manner.

Hardware takes on the complexity. With the engraving techniques available today and the exhaustive testing methods, we can repeat the functional patterns of informational transducers, reproduced identically, as many times as the surface of used silicon will allow. Validation is 100% under control and circuits are monitored every hour of the day, every day of the week, in real time.

Software takes care of the rest, thus respecting the Turing principle of keeping "the hardware as simple as possible", since everything can be known through the construction. The fundamental virtue of software is that it is modifiable and adaptable, without any material constraints, which is not the case for hardware. We can thus develop statistical approaches, draw curves representing aging, which would be incompatible with the specific constraints of the hardware, and fully exercise the PAC logic defined by Leslie Valiant[35].

To try to have the hardware play the same role as the software is thus a total oxymoron. It moreover does not prevent us from transforming completely stabilized software functions into hardware, since each remains in the role where it is optimum; that is the task attributed to tools such as the silicon compilers that simultaneously generate the work to be done and the validation of this work. It is

35 See his book, *Probably Approximately Correct: Nature's Algorithms for Learning and Prospering in a Complex World*, 2013; a logic determined by the third principle of systemics (Printz 2019).

also the objective of a language such as SDL, used for protocols, and which generated the language of testing, TTCN, according to the specifications written in SDL. The possibility of unknowns given the combinatorics we are now faced with in current computers makes it indispensable that the nominal computer be monitored by another computer exclusively devoted to this surveillance. This makes for a functional redundancy, and we can progressively introduce all the know-how of systems architects and users into the surveillance computer depending on the environments it is used in. This is the principle of what is called an autonomic cycle, which was discussed in the previous chapters. It is the same as saying that in this type of architecture, everyone monitors everyone else, following the negotiated protocols. There are no longer centralized "watchmen" in the sense of von Neumann's question: "who guards the guardians?". The assurance of proper functioning is an overall property that every part contributes to, depending on the level it occupies in the functional hierarchy defined by the stack, which itself accounts, since the start, for the criterion of testability.

If we look at the interoperability diagrams used in systemics through the lens of a computer stack, we note the fundamental role of this logical structure even more; its exceptional virtues owe nothing to chance but everything to the intelligence of its creators[36]. The interaction between systems is first and foremost an interaction between the stacks of these systems that manage the inside/outside relationship, stacks which must therefore be logically and semantically coherent among themselves (see Figure 3.15).

Systems have maximum autonomy when the level of interaction is at a higher level in the stack. We can also understand that to interoperate in a secure manner, it is important to respect the logic of levels. Connecting different levels together is a way of managing events that the superior levels are unable to interpret, and inversely, as in Figures 3.5 and 3.8. We note that the more interaction is completed at the lower levels of the stack, the more the systems become rigid and synchronized. The more we rise in abstraction, the more the systems gain in autonomy and cooperative capacities, which makes their robustness and resilience better, but on the express condition that the logic and the semantics of the exchange model are scrupulously respected.

The computer stack is the materialization and the translation in language of systems engineering from the conclusion of our text *System Architecture and Complexity: Contribution of Systems of Systems to Systems Thinking*. It perfectly illustrates how to implement the three laws of systemics defined in this book, as well as the role delegated to the architecture, recalled in Chapter 1 of this book.

36 See Chapters 8 and 9 in Printz (2019) as well as Chapter 1 of this book and Figure 1.3.

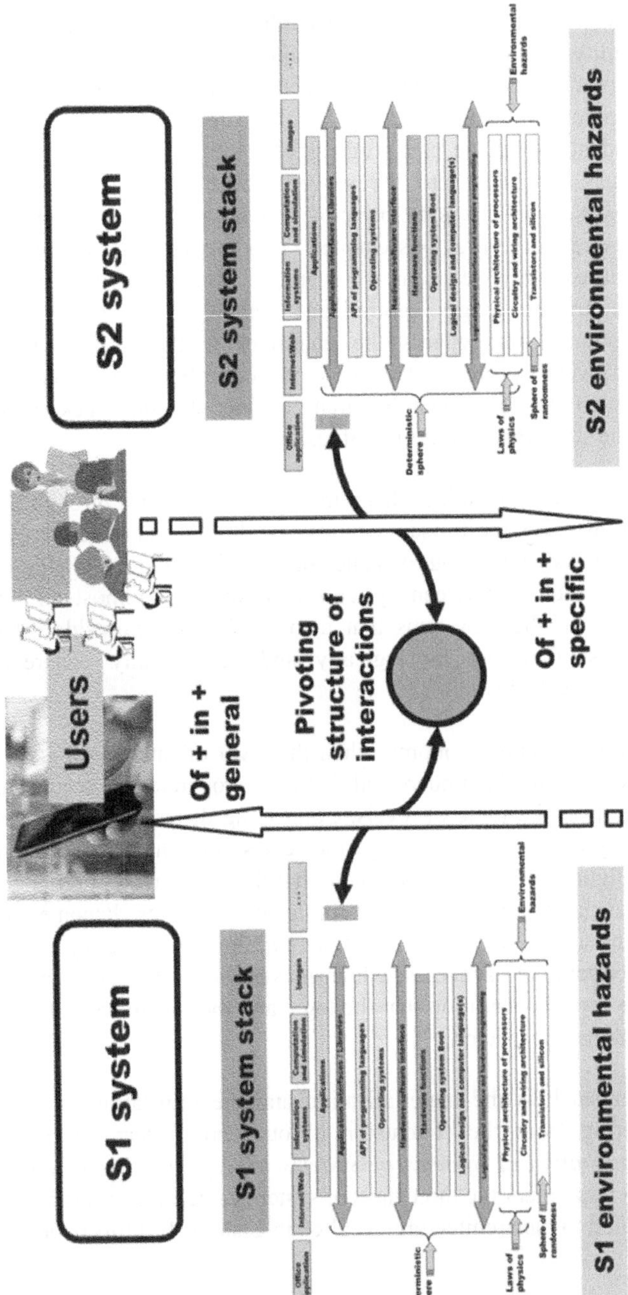

Figure 3.15. *The computer stack and interoperability*

PART 2

A Complex World

Introduction to Part 2

Complexity has become the number one problem for all our present and future systems.

Speaking about "complex systems" has become a pleonasm. Just as chaos is at the very heart of rational mechanics, complexity is present from the start, at the level of the hardware and software components throughout the whole system and in all of their interactions, even though the studies initially conducted by von Neumann on the hardware component were completely general. Software – this new somatic form also generically called information engineering – had adapted to this, except in so-called "real-time" system software systems and in operating systems, but the engineering communities in charge of them came together as early as the 1960s to address this, leading to grandiose developments such as the "computer stack" presented in Part 1, or great technological systems such as the electrical system, which will be presented in Part 3.

After the widespread computing revolution, it is now the entire engineering community, with all the disciplines intertwined, that is faced with the issue of complexity. The INCOSE report, "A World in Motion: Systems Engineering Vision 2035", is very clear regarding this ongoing subject, which will remain so for the next several decades[1]. Everything is digital now: organic analog equipment to control the regulation of systems has virtually disappeared, being replaced by functionally discreet systems, that is, computers containing billions of components. There has been a change of scale, of at least two to three orders of magnitude. Engineers faced with programming in the larger sense, in other words the information processing that is necessary for the systems, now number in their millions; the creation of a 3D printer, for example, is more than anything a feat of programming. The self-driving car and the 100 million lines of code needed for its integration into urban traffic

1 The report can be downloaded from the INCOSE website.

reveal the true face of the complexity: a risk and an opportunity, together. It is therefore primordial to have sound ideas on the subject of complexity, to know what we can do and to know the limits/limitations.

Systemics is to this day the best way, using models, to formalize this approach in a constructive way and to manage the rise in age of engineering teams dealing with complexity. If this is not the case, the INCOSE's discussions about a World in Motion will be nothing but wishful thinking.

In this part of the book, we will present an aspect of complexity that has until now been rarely discussed, but that represents the everyday experience of systems architects and directors of systems projects that are called "complex" due to their size, cost, technical and social risks, the number of interested parties interacting to define them and the size of the development teams needed to build them. For them, at one moment or another, complexity will mean a volume of verification and validation work that is either accessible or not, given the economic conditions that we live in, so they can satisfy the external constraints of the system; and sometimes it will even mean the impossible, when they are faced with the intrinsic limitations that only really appear in information science.

4

Phenomenology of Complexity

For the last 20 years or so, each of us, at least in societies which have benefited from the advantages of globalization, has had the feeling that we are living in a society whose complexity, in the basic sense of the term, is accelerating and perhaps surpassing what we are able to comprehend, especially given the average age in these societies along with the maturity of the individuals making them up (what is called the "grade", coming from the word "graduate", both on the individual level and the collective). In China, also in full acceleration, things are certainly different than in our Western societies[1], which are getting a bit worn out, and starting to decelerate; they did however propel this wave with the Industrial Revolution of the 18th and 19th centuries.

In the INCOSE document, *SE Vision 2035*, fields in complexity science were central, which could also be seen at the Complex Systems Design and Management (CSDM) conference, presided over by Professor Krob, over a period of more than 10 years now. Complexity is at the heart of artificial information and science, to paraphrase Simon. This is also what we will illustrate in Figure 4.1.

In this figure, we see the three complexities: the users/customers of the system (U), the organic elements of the system (S) and the systems engineering project and its actors (I), whom we must imagine to be in constant interaction throughout the entire life cycle of the system, which can stretch out over several decades.

1 von Neumann, in a remark reported by his friend Stanislas Ulam, is reported to have said: "The interests of humanity might change, the present curiosities in science may cease, and entirely different things may occupy the human mind in the future." We are almost there when we see the lack of interest in scientific training, despite the fact that our society is totally dependent on science and mathematics.

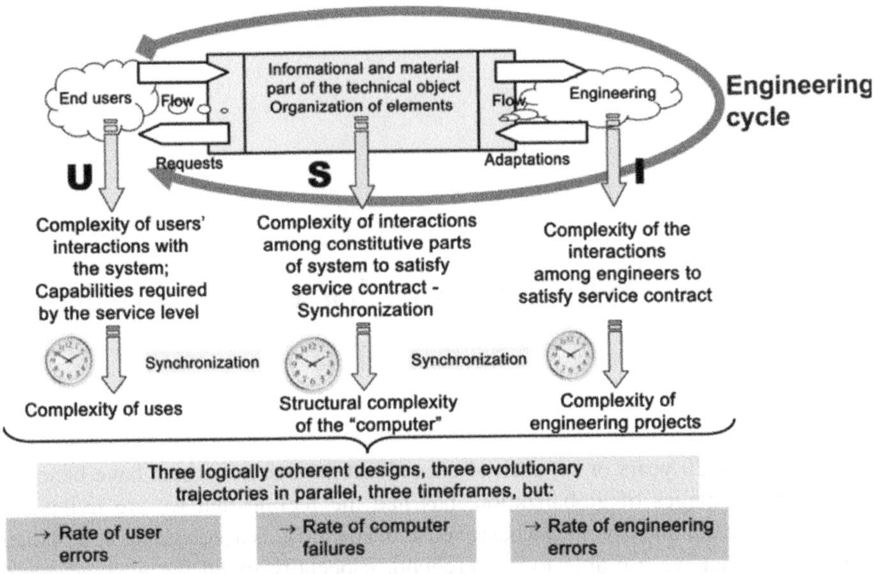

Figure 4.1. *Complexity in interactions. For a color version this figure, see www.iste.co.uk/printz/pedagogy.zip*

4.1. Drive and control in a complex environment

The word complexity only began to appear in scientific vocabulary in the 1950s with the arrival of computers, and thus von Neumann's "computing instruments" was considered to be entirely impossible at the time for reasons of reliability. At the time, von Neumann himself spoke of "complexity", "complication" or even "ultra-high complication" interchangeably. Von Neumann played a major role in the quantification of this realization, in particular, via the theory of automata, which motivated him throughout the entire last decade before his untimely death; at the age of 53, he passed away from bone cancer which he had probably contracted in the first nuclear test sites for atomic bomb experiments. In epistemology, it was not until the 1970s and 1980s that the word complexity began to appear. This story was told over the course of a series of conferences at the Université Catholique de Lille in 2017–2018[2].

The complexity we see can be analyzed by extracting a four-dimensional reference framework which will enable us to capture all of the nuances.

---

2 Published as *Survivre à la technologie ? – Aux sources du cyberespace et des sciences de la complexité* with Saint Léger Editions in 2018.

Firstly, there is static complexity, which is only a simple nomenclature for the constitutive parts of a computer or system, pieces which can themselves be complete systems when viewed from outside, like indivisible monads, from which we get terms such as "black boxes" to designate the terminal elements of hierarchies seen with configuration management tools in systems engineering. These descriptions are normalized. This is basic complexity, sometimes called "complication" by some. It is a simple calculation for which we must nevertheless clearly define the counting rules in order to avoid confusing everything.

Next, there is dynamic complexity, which is made of relationships that develop over the course of the lifetime of a system in its usage/operation phase. This dimension can be seen in mechanisms such as the so-called N2 traceability matrices, well known in systems engineering. On the one hand, there are the functional coupling matrices between the functions themselves, and the functional dependencies that result from those which are near each other, thus forming significant chains of hierarchical functions, also called "lifelines" – that is, "sentences" in the linguistic sense of the term, and for which the system constitutes the grammar – and coupling matrices between the functions and the states at which they operate, called state/transition systems in the theory of robotics, following the classic model $State_1 \xrightarrow{Transition} State_2$. The representation of these mechanisms has existed from the start with notations such as SADT/IDEF, and more recently, UML/SysML, as recalled in Chapter 1. In the end, we can set out the frequencies of states and transitions for a given execution by the system over a given period of time with an established reference "load", called a benchmark in jargon.

The notion of an automaton as formalized by von Neumann in his foundational studies takes this double aspect, static and dynamic, into account, mentioning the "complication", or even the "extremely high complication" in large automata[3]. It is at the heart of these surprising computers that came into being in the 1950s.

4.2. Communicating in a complex environment

Another aspect of complexity appeared with the technology needed to exchange information between the thousands of active elements in the first computers; the Whirlwind computer of the SAGE project contains about 30,000. Not only can the communication be "noisy", which can compromise the quality of the transmitted signal – something we have known since the first work on telephone lines which began to be installed almost everywhere, as well as on radio emitters/receptors – but

3 The idea had been floating around since the 1930s, in particular with Gödel, and especially Turing with his *Logical machine*, which would later be called the *Turing Machine*.

the constitutive elements themselves can be the source of random failures, meaning the failure of just one part can compromise the correct functioning of the whole.

The world of these new computers is a world governed by the laws of probability, such as those which govern the behavior of atoms and molecules, or even those which appear to govern the behavior of socio-economic actors, whom von Neumann modeled in his foundational work, *Theory of Games and Economic Behavior* (1944), a book which enjoyed a glorious posterity at the RAND Corp. in terms of Nobel prizes.

Shannon published article that made history in the *Bell Labs Technical Journal* in 1948, revisited in the book he published with Weaver, *The Mathematical Theory of Communication*, in 1949. He showed the central role of the probabilistic approach, not only in the construction of economic codes but also for the construction of redundant codes used for compensating the vagaries of "noise", codes that were for this reason called "error correction codes", without which none of our computers would work for more than five minutes without crashing.

In summary, uncertainty, randomness and even the unknown are everywhere in these computers, and not addressing it makes a computer impossible to make because reliability and functional assurance cannot be controlled. As Hamming said, taking care of "noise" and "errors" is a central objective of information sciences.

What is, in a certain way, "magical" in random phenomena is that despite intrinsic ignorance about inaccessible individual behaviors, there is an emergence of collective behaviors that are perfectly stable and deterministic, which all the physics of solids – that of condensed matter – bears witness to. This physics is that of now nanometric components, at the heart of all our computing systems, and which are subject to the randomness of the quantum sphere. It is also what the logic of the computer stack manifestly shows, since without it, no reliable communication between computers would be possible, including within its own programming.

Neither ignorance nor – much less – incertitude are obstacles to making these fantastic computers, as long as we go about it in the right way. That is how von Neumann spent the last 10 years of his life, studying the theory of automata; I am specifically thinking here of his 1952 essay, *Probabilistic Logics and the Synthesis of Reliable Organisms from Unreliable Components*, and the works that followed such as those by Winograd and Cowan, *Reliable Computation in the Presence of Noise* (MIT Press, 1963) among others.

We do know the answer to these truly fundamental questions for the future of these technologies: it is organization, or architecture, as has been said since the 1960s in the field of this new type of engineering that has kept us from falling into the hell of combinatorics, which we will discuss later on. A kind of engineering that leans on a fundamental science which helps us understand the ins and outs of what we habitually call "complexity", but above all, helps us to adapt to it and even take advantage of it in the engineering of systems so that we no longer even call it complex since it is a pleonasm. The dynamics of any system, because of the feedback and the hazards of the environment, is intrinsically complex.

Leslie Valiant, a computer scientist specialized in telecommunications and recipient of a Turing Award, in his remarkable book, *Probably Approximately Correct*, as cited above, and in which he describes the logic at work in all computerized systems, gives us a precise operational characterization of these somewhat qualitative notions that paved the way for action.

Uncertainty is characterized by what he calls *theory-full*, that is to say that we have a theory, like, for example, statistical physics or quantum mechanics, or the theory of queues, which explains the random, observed phenomena by calling on probabilities to explore the resulting combinatorics. The performance of systems derives from this approach, including even the determinist chaos that comes from simple laws[4]. Control of the tasks in our everyday systems also arises from this.

In the contrary situation, ignorance, when we are faced with an unknown phenomenon whose laws, by definition, we do not know, we are what is called *theory-less*, that is, we are not entirely helpless, since through simple observation and collection of information, we can construct an empirical model of a phenomenon, and of the limits of its validity, above all; this model does not have the purpose of "explaining" the phenomenon, but just describing it case by case. This description corresponds to what, in group theory, is called an extension. In systems engineering, monitoring systems like capacity planning (CP) all fall into this category. In asynchronous communications between systems, of which the Internet is the emblematic example, we often do not understand the operations in detail, but we know how to take control in the case of damage to restore the service.

This is a distinction between theoretical models on the one hand and empirical models on the other, which has been part and parcel of modern science and its methods since Galileo or Newton. It was the subject of a presentation by von Neumann given for the bicentennial of Columbia University in New York in 1954 during a conference on the unity of science called *Method in the Physical Sciences*, reprinted in the *Complete Works*, vol. VI.

---

4 See the article by May, R. (1976). Simple mathematical model with very complicated dynamics. *Nature*, which gives an excellent example of this; see also Ruelle, D. (1991). *Hasard et chaos*. Odile Jacob.

The "end of certitude", to paraphrase the title of a book by Ilya Prigogine, does not mean that we have reached our limits; it means we must approach problems differently, and that we must really understand what makes us say that something is "impossible", and that such a situation will "never" occur. In engineering sciences, as in experimental physics, we learn to ignore certain parts of phenomena that interest us, what we sometimes call "line thickness" or the randomness that is used to designate small quantities, as long as the laws and/or rules are known, and especially as long as perspective is not lost. Mathematicians have taught us that a sum of infinitely small quantities can itself be infinitely small, but that sometimes it diverges, either towards a finite number or towards infinity; this is the limit of engineering that we will examine in terms of combinatorics in this second part of this book.

4.3. The four dimensions of complexity

The reference framework for complexity imagined in this book therefore has four axes, that is, is measured in four dimensions in the sense in which this term is understood in physics with so-called dimension equations – for example, energy has the dimensions $M \times L^2 \times T^{-2}$ in which mass, distance and time are taken into account, and entropy has the dimension of energy divided by °K, which makes very useful qualitative arguments possible – or, for complexity as we understand it:

– complication, that is, static complexity, as a simple count of the elements that can nevertheless be arranged in a hierarchy by levels of composition and/or abstraction;

– complexity, a term that we use when speaking about dynamic complexity, that is, the combinatorics created by coupling matrices and exchange channels throughout the life of a system;

– uncertainty, that is, the arrangement of a theoretical model of one or more phenomena occurring in a system that allows for different simulations/predictions that help understand the reality; this is Valiant's *theory-full*;

– ignorance, that is, the arrangement of a descriptive, empirical static model, on a case-by-case basis, that nevertheless allows us to identify verifiable limits and/or constraints or to raise alarms; this is Valiant's *theory-less*[5].

Axes 1 and 2 represent static and dynamic combinatorics, in other words, that which is stable in the general process or, if you will, in the counting of elements and the interactions between them that the system brings about. Axes 3 and 4 reflect the

5 In the modeling languages currently available, this is the model for where-used lists or "use cases".

state of limited knowledge that engineering teams have access to for resolving the problems that present themselves, however well-educated they might be, individually and as a whole; that is, what is instable in the general process but that can nevertheless be controlled. Axis 4 reflects an essential quality in engineers: common sense and intuition, knowledge of the "field", and both von Neumann and Turing underscored the importance of this[6]. This division over 4 axes is also in keeping with the analyses by Forrester (1961) in his foundational book, *Industrial Dynamics* (MIT Press), which guided the entire discipline. In fact, this is a continuum that systems engineers are presented with.

We should remark that the simplest complexity is that of a structure of N linearly connected elements, without cycles, or exact $N-1$ relations, which could possibly be useful in defining a measurement standard.

As has been shown in other books, in systems engineering, complexity is measured in energy, that is, for the project manager, with a cost, estimated in hours worked, or man-hours (mh). In physics, we know that energy can be broken down into kinetic energy, potential energy, thermal energy, etc., all linked together by the laws of conservation of energy. We see analogous phenomena in the models for costs, since, like "energies", complexities are linked to each other through information – an analogy that it would be useful to analyze fully.

In the system reference framework, what simultaneously defines the *what* and the *how* – the objective and the method to reach this – will automatically create a cost, or rather, a portfolio of costs in which the system being made can be found in the end. Let us note that the absence of a reference framework constitutes in itself an implicit reference framework from which we incidentally can draw no conclusions; it is a dive into the unknown.

We can invert the perspective and ask which system, with which functions and what level of service can we obtain for such and such a portfolio of costs, regardless of the level of the actors, users and engineers, or the clients who will pay the bills at the end. This is what is known as affordability.

We must thus project complexity along four axes to determine the modalities of the actions to follow through on in each direction. For example, entirely new systems, such as SAGE and NTDS when they were in their heyday, require a large volume of fundamental R&D – axes 3 and 4 – which, at the time, motivated the

---

6 According to Dyson, one of von Neumann's most reliable biographers, "All processes that are stable we shall predict. All processes that are unstable we shall control"; this is the basic hypothesis of the foundations of cybernetics (see Printz 2019), and systems architecture, recalled in Chapter 1. But for this, we must first be well trained.

teams at MIT, Princeton and some of the other top American universities. As information science has clearly demonstrated, axis 3 conditions what happens along axes 1 and 2.

For example, not knowing the theory of automata to address certain combinatorial problems such as developing compilers for high-level languages will automatically cause an inflation in the number of lines of code in a report, which can go from 20 to 100, and which may mean that the corresponding program is impossible to make[7]. It is the same for telecommunications protocols such as TCP/IP and http, which are simply impossible without knowing the theory in detail, or having a management systems for databases which integrate a variety of unknown but indispensable formal languages used by creators.

4.4. Measuring complexity

The approach recommended by von Neumann, via the theory of automata, created families of measures.

The first family consists of counting the number of steps in the calculation that needs to be done to obtain the result – that is, the number of transitions/transformations needed to go from the initial state to the final state of the automaton, resolving the problem it poses.

This measure of the dynamic dimension gave rise to the theory of algorithmic complexity, and even that of depth, by Bennett. Von Neumann's initial motivation was to create rapid calculators to get through the volume of calculations needed for the study of H bombs or meteorology, via simulation. Algorithmic complexity is an important chapter of theoretical programming; that is, what made it possible to obtain the number of operations needed based on data. For example, a sorting operation of N elements requires a number of operations, on average, $O(N \times \log(N))$, which is read as "of the order of magnitude of...". These formulas are primordial for analyzing the performance of systems and eventually their scalability.

The second family concerns the necessary and sufficient size of programs required for the resolution of a problem using an algorithm defined on a standard

7 As a reminder, a compiler node in a high-level language such as C/C++, Java and Python contains about 120,000 strongly coupled lines of code (compilers are often "bootstrapped" in their own language), or a text equivalent to six 400-page books; a complete compiler with all its appendices is equivalent to about 250,000 to 300,000 lines of code. The first compiler, Ada, by Jean Ichbiah from the company Alsys, and for which I was one of the reviewers, contained approximately 350,000 lines of code, written and "bootstrapped" in Ada.

computer, that is, the number of successive decisions/choices the programmer needs to make in order to put their program together.

Choices are made from a set of possible cases that we are sure about, the basic commands of the computer – its internal language, usually hidden, based on the transducers it makes available to us – whose composition is the program that results from these choices. In the end, it is the size of the calculation blueprint in the corresponding Turing machine, represented by a matrix, that becomes with a few basic commands a measurement standard for textual complexity; these are the measurements of the static dimension.

What is interesting in these measurements is that they are directly correlated to the work of programmers. Any act of programming, like configuring a system, in the largest sense of the term, is the product of a measurement of textual complexity that we can try to minimize by working on the architecture of the materialized text using one or more configuration files[8].

The steps in building/taking apart any device, whether it is an automotive vehicle, an Airbus A380 or a smartphone, is also a product of this approach[9].

We should note that the richer, more diversified and adapted a language is to a situation, the shorter the resulting text will be, but the more difficult it will be to acquire and master it, so it will increase the rate of errors among its users. There is necessarily a compromise, which can vary with the situation and/or environment; nothing is ever definitive.

A validated/tested program works, for the user, like a theorem that they can trust. But the theorem/program must still be proven to conform to the accepted rules of the community of mathematicians/programmers. The size of the proof must therefore be integrated in the measurement, that is, the size of the testing programs. The size of these programs, which reflect the activity of IVVT (integration, verification, validation and test), a set of fundamental processes in systems engineering, accounts for the dynamic dimension of complexity, which means that the sum of these two measurements gives a good idea of this static and dynamic complexity[10].

---

8 See Printz (2013a, 2013b).

9 A colleague at Renault pointed out to me that recent Formula 1 cars are made up of 37,000 pieces; much less than a smartphone with its billions of transistors and its tens of millions of lines of code!

10 See my lecture at the CSDM Conference in 2010, *A Natural Measure for Denoting Software System Complexity*.

With only this reservation, the measuring of textual complexity, whose origins go back to the work of the mathematician Andrei Kolmogorov, completed by that of Gregory Chaitin – what has since been called the *algorithmic theory of information* – is an indispensable tool of metrology for the systems architect and project manager, as are the most basic tools and methods such as nomenclatures – also called ontologies – and matrices of functional dependencies like the N2 matrices of systems engineering, which are all static measurements.

4.4.1. *On Richard Feynman's intuition*

In his Nobel Prize lecture of 1966, Richard Feynman made an interesting remark that clearly illustrated his pedagogical genius as much as his creativity, outside of what was in fashion and the previously taken paths. He told us, simply:

> It always seems odd to me that the fundamental laws of physics, when discovered, can appear in so many different forms that are not apparently identical at first, but, with a little mathematical fiddling you can show the relationship. [...] Perhaps a thing is simple if you can describe it fully in several different ways without immediately knowing that you are describing the same thing[11].

He tells us that it is perhaps a way of defining simplicity, which is another way of defining complexity, by taking the inverse: what is not simple is complex; what can be described in different ways is simple.

Besides programming, which offers, through the multiple languages and/or equivalent representations that is used regularly, innumerable examples of this diversity of textual formulations for the same algorithmic transformation, mathematics itself and more especially geometry, can serve to illustrate the depth of Feynman's intuition; he would also say "What I cannot create, I do not understand." This constructive way of defining complexity is compatible with the textual complexity of Kolmogorov/Chaitin since a "good" language must reduce textual complexity.

It is also the case in our example of the computer stack, which uses a great variety of languages, from the silicon to the user interfaces whose "simplicity" is merely apparent, since the interface stack actually masks a formidable complexity that is found in the engineering costs of these interfaces, which cost tens of millions of working hours. The interface language makes the underlying complexity "simple". This stack illustrates Turing's intuition, well before it appeared in real

11 See: https://www.nobelprize.org/prizes/physics/1965/feynman/lecture/.

computers, presented in a little-known article, "Systems of logic based on ordinals", written in 1938 during his stay in Princeton where he finished his thesis with Church.

In modeling, we most often begin by describing what we want to model on a case-by-case basis (we use the expression "use cases", which is a representation by extension, understandable to users), before proceeding to a generalization in an algorithmic form when possible; the result is a representation by intension, more compact (think of the gains obtained when we go from the voluminous numeric tables used by engineers before the era of calculators, to algorithmic representations, available today in our smartphones; Laplace would say that the invention of logarithms had doubled the useful life of astronomers).

In the theory of formal languages, there are equivalence theorems related to the description of a language, either in the form of a grammar (external language, for users, defined by its grammar) or in the form of automata (internal language, for engineering). From an energetic point of view, a text in grammatical form is much more economical than the equivalent text compiled in the form of an automaton expressed in the language of the computer, in a ratio that can be of the order of 1 to 100 (two orders of magnitude). Hence, the interest in and the economic importance of these equivalences in systems engineering since the scope is such that we can go from doable to impossible. If the architect does not "see" the abstractions that underlie the case-by-case formulation, they will miss the simplicity, in Feynman's sense. We go from the grammatical form, the most abstract, to the automaton form, more concrete, through compilation techniques that are now completely controllable; the same notion exists in the duality of functional language/imperative language. This is a practical application of Hilbert's motto "generalize to simplify".

In geometry, we know the importance of the choice of one coordinate system or another (Cartesian, polar, affine, intrinsic, etc.), called the reference framework, for resolving a geometric problem, or even the choice of the most appropriate geometry, be it Euclidian or not, with its specific axioms, as Poincaré pointed out. For example, the classic geometry of cones, using a ruler and compass, is drastically simplified when we formulate it in terms of projective geometry, a step in the evolution towards Grothendieck's algebraic geometry.

With information engineering, what plays the role of the reference, or the coordinate system, is the underlying abstract computer(s) and their expressive capacity (operational semantics, defined by the available transformation mechanisms on the computer, including the available API libraries), that is, their instruction manuals, including everything up to the transducers, which in the end will provide the service requested of the system in the spatiotemporal physical reality of its real environment, integrating the corresponding hazards, including the errors and the "noise" of the environment. This is what we have called the metamodel, or more simply, a second-level model.

The multiplicity of languages associated with these virtual computers, however well-targeted they may be, is clearly a factor of simplicity, in Feynman's sense, which can appear paradoxical and/or counterintuitive, but this is what the example of the computer stack demonstrates. We must not be mistaken however: the languages in question are in a complementary relationship with themselves, the one doing what another cannot do, or does badly (e.g. the data language SQL or the Web communication protocol language HTTP, or the one defined in a DoD/NATO norm for GUI defense and security systems)[12].

Conversely, the general languages, Java, Python and C#, for example, do not have this type of relationship; they are simply equivalent, and only the available API libraries could perhaps distinguish the best one. Using them without reason within one single application means increasing its complexity, not reducing it.

Creating languages adapted to a situation is a factor of simplicity, and yet we must have the ability to do this, which corresponds to axes 3 and 4 in terms of knowing language theory and compilation techniques, disciplines which we should point out have not been taught in France since the 1990s, for almost a generation.

4.5. Counting

Learning to count is the first step in analyzing complexity as proposed in this book; it is the first dimension. If we take Simondon's triplet {U, S, I}, we must imperatively have a clear idea of 1) the number of system users; 2) the number of constitutive elements in the system; and 3) the number of project actors for the system, that is, all those who participate in its engineering to any extent, and who are perhaps also users.

This basic counting depends on the point of view that is adopted, which we understand better with information science. In mechanical objects/systems, such as a watch with complications or a Formula 1 car, we count the pieces, or the teeth in the gears, not the atoms or the molecules; we account for phenomena of aging, failure rate, etc.

12 *Common warfighting symbology*. See: http://everyspec.com/MIL-STD/MIL-STD-2000-2999/MIL-STD-2525_20727/.

In a programmed object/system, it is different since the physical systems housing the information, the nanometric components, are sensitive to quantum noise; consequently, how we count can be a real problem.

What is subject to corruption are the programs, that is, the instructions and the data manipulated by the instructions, for which we must know the volume and have as precise an idea as possible of the rate of residual errors. There are counting rules that were established by the Software Engineering Institute (SEI) at CMU that we will not recall here, as what matters is to point out that the interface determines the counting boundary: what is below, the internal language (IL) in von Neumann's terminology, and what is above, the external language (EL), accessible to users. Insofar as the interface is a hermetic boundary, what counts is the size expressed in the counting rules of the EL as we go through the hierarchy of abstractions defined by the architect to the root of the system.

We can measure the effect by examining the size of the N2 matrices. The 100 million lines of code cited for self-driving vehicles become approximately 500 million computer instructions. If the interaction happens at the level of computer instructions, that leads us to completely unmanageable matrices of $(5 \times 10^8)^2$. If we reason by correctly ordered and autonomous blocks of modular code, let us say on average 20,000 lines of code per block, the size of the corresponding N2 matrix is now only $(5 \times 10^3)^2$, which is manageable both in terms of human capacity, from a cognitive point of view and from the point of view of the technology that is currently available and/or will be in the next 15 or 20 years.

This is what we will now examine in detail, since combinatorics are a fundamental and principal aspect of complexity, by reflecting on the numbers qualified as inaccessible by the mathematician Émile Borel.

5

The Roots of Complexity: Inaccessible Numbers

5.1. The provenance of inaccessible numbers

Immense numbers that defy the imagination, as Boltzmann used to say, made their entry into the world of mathematical physics with atomic theory and thermodynamics in the form of the kinetic theory of gases, which was developed by physicists like Maxwell, Boltzmann and Gibbs towards the end of the 19th century.

The flagship number in this microscopic world that is inaccessible to our meso-macroscopic senses is the Avogadro constant, abbreviated N_A, after the publication of Amedeo Avogadro's thesis, *Essai d'une manière de déterminer les masses relatives des molécules élémentaires des corps, et les proportions dans lesquelles elles entrent dans ces combinaisons* ("Essay on a manner of determining the relative masses of the elementary molecules of bodies and the proportions by which they enter these combinations")[1].

The atomic hypothesis was only taken up a century later, and we can consider the publication of Jean Perrin's book, *Les atomes*, in 1913 to be the final, settled account. This book will always be reedited[2].

Its current value is estimated as 6.02214076... × 10^{23} per mole, valid for any simple entity, for example, water, whose composition is H_20, or one atom of oxygen

---

For a color version of all the figures in this chapter, see www.iste.co.uk/printz/pedagogy.zip.

1 *Journal de Physique, de Chimie, d'Histoire naturelle et des arts*, July 1811, 73, 58–76; see https://en.wikipedia.org/wiki/Amedeo_Avogadro.

2 Currently in the Champs collection with Flammarion, with a preface by Pierre-Gilles de Gennes.

and two atoms of hydrogen for one molecule of water. Note that despite the precision obtained – nine significant figures – incertitude shrouds 15 digits, on the order of a million billion molecules.

Today, thanks to X-ray analyses, we know the form and size of a water molecule. During Avogadro's time, the only things known for certain were the respective proportions of hydrogen and oxygen, following the chemical reaction $2H_2 + O_2 \rightarrow 2H_2O$, which can be read as: two hydrogen molecules, or 4 grams, combined with one molecule of oxygen, or 32 grams, to form two molecules of water, or 36 grams, that is, 2 × 18 grams, or even two moles. The constant N_A lets us go from the microscopic world of atoms and molecules to the meso-macroscopic world where we can measure weight using scales, as Lavoisier did in his quantitative analyses of combustion. Hence, the interpretation in terms of proportions, indispensable for chemical engineers, namely, {4, 32, 36}, which can even be expressed by taking the lightest entity among them, hydrogen, as a unit {1, 8, 9}, introducing the notion of molecule/mole, or just the mole[3]. The delicate constitution of atoms and molecules was not understood until much later, with atomic notations that are today well-known, distinguishing the protons, neutrons and electrons, as the mass is essentially concentrated in the protons and neutrons, commonly called nucleons, denoted as ${}_{Z}^{A}X$ in which X indicates the entity, A the mass (number of nucleons) and Z the atomic number (number of protons, and therefore the number of electrons)[4]. For water, we have the hydrogen molecule, composed of two hydrogen atoms, or ${}_{1}^{1}H$ (only one proton), the oxygen molecule, also composed of two atoms, or ${}_{8}^{16}O$ (but the oxygen we breathe also contains isotopes of a mass of 17 and 18, although in very small proportions, hence an equilibrium for the macroscopic mass), and the water molecule, formed of one atom of oxygen and two atoms of hydrogen[5].

In the liquid state, this means that 18 grams of water, or a small glass of 18 cl, contains approximately N_A molecules of water. In the gaseous state, in the form of vapor, with a pressure and boiling point under 1 bar, 1 liter of water, or 1 kg or 1,000 grams, will produce 1.694 m^3 of vapor, that is, for our 18 grams of water, a volume of $18 \times 1.694 \cong 30$ liters[6]. This will be the same in the solid state, where ice, which is a little less dense than water, will float, as all of us can see; which means that there is more empty space in ice than in water.

3 See: https://en.wikipedia.org/wiki/Mole_(unit).

4 See the Periodic Table of the Elements: https://en.wikipedia.org/wiki/Periodic_table; the mass of an electron is 1.836 times smaller than that of a proton, which is negligible in an estimate.

5 See: https://en.wikipedia.org/wiki/Properties_of_water.

6 See: https://www.thermexcel.com/english/index.htm and the correspondence tables for T and P.

This number, N_A, expressed in common language, can be presented as 600 thousand billion billion, recalling that 1 billion is written as 10^9 in the notation of engineers. We can compare it to the number of neurons in the human brain, evaluated at 100 billion, or 100×10^9, or perhaps to the set of cells in an adult human body, something like 100 thousand billion cells, or 10^{14}, with all categories combined (there are a little over 200). There are therefore many more atoms/molecules than cells. Given that the human body, and more generally, living matter, is essentially composed of water, we could calculate an approximate ratio, given that 1 liter of water (1,000 grams, or $\frac{1000}{18} \times 6.02 = 334{,}444...$) equals approximately 335×10^{23} molecules of H_2O.

In what we call nanotechnology, we are at the atomic scale, since 1 nanometer is equal to 1 millionth of a meter, that is, in the notation of engineers, 10^{-9} meters. Since the 1980s, we have known how to operate, under certain conditions, at this scale thanks to the scanning tunneling microscope, invented by researchers at the Vienna IBM laboratory[7]. The fineness of electronic circuit engraving in the silicon of our computers is currently on the order of 10–20 nanometers, using techniques from lithography with electron rays, the equivalent of a row of about 100 silicon atoms[8].

5.2. Typology of inaccessible numbers

There are two sorts of huge numbers. There are those which we just discussed, grasped by physicists with the help of probability calculations, a field that developed remarkably over the course of the 19th century, in particular with Laplace and Gauss, and which became the language of statistical physics, entirely created by Maxwell and Boltzmann[9]. This first category of numbers are those of our physical world, as we can grasp them using our senses and our observation instruments, themselves part of this world. With the demonstration of atoms, on the one hand, Planck's discovery of how to quantify energy in reactions, and the start of quantum mechanics, on the other, the language of probabilities began to seem like the ideal instrument for describing phenomena on our scale such as temperature, pressure and/or viscosity, which are the macroscopic result of the "typical" behavior that groups of atoms like those we just discussed exhibit. Manipulating this language is

---

7 See: https://en.wikipedia.org/wiki/Scanning_tunneling_microscope.

8 See: https://en.wikipedia.org/wiki/Electron-beam_lithography.

9 See Roger Balian's lecture at the Ecole Polytechnique: https://en.wikipedia.org/wiki/Roger_Balian and his monograph, *Une crise en physique? Entre déterminisme et imprévisibilité*, available at: http://ipht.cea.fr/Docspht//articles/t09/153/public/Lille_crise_doc.pdf.

difficult, often subtle, and always counter-intuitive (see section 6.1.1.1). Hilbert, in problem number 6 of his *Mathematical Treatment of the Axioms of Physics*, said[10]:

> Investigations into the foundations of geometry suggest the problem: treating the physical sciences, for which mathematics already play an important role, in the same way, through axioms; the theory of probability and mechanics come first.

He explained:

> Regarding the axioms of the theory of probabilities, it seems desirable to me that logical investigations be accompanied by a rigorous and satisfactory development of finite gains in mathematical physics, in particular in the kinetic theory of gases. […] Boltzmann's work on the principles of mechanics [NB: what will become statistical mechanics] suggests the problem of mathematically developing the limiting processes, only just outlined, which lead from the atomic perspective to the laws of continuous motion.

The fundamental problem was how to represent discreet phenomena, atoms or quantum energy, with the mathematics of continuity, without inadvertently introducing an interpretation related to physical phenomena (see the quote by Feynman in section 6.4.1).

But soon, a new category of immense numbers appeared, those which we qualify as infinite complexity or infinitely complex, from the sphere of information whose basis lies in the complexity of life, especially with the arrival of the first computers in the 1950s and the discovery of DNA.

Figure 5.1, taken (and altered) from the book by Gilles Cohen-Tannoudji (1989), *L'horizon des particules – Complexité et élémentarité dans l'univers quantique*, (Gallimard, 1989), perfectly summarizes the current situation. This is a book about the history of elementary particles, comparable to Weingerg, S., *The First Three Minutes*, first edited in 1977, and most recently in 1993.

Beneath the level of the elementality of the particles that make atoms, there is another level, which is the standard model of particles, formulated from the 1930s

10 See the 23 problems: https://en.wikipedia.org/wiki/Hilbert%27s_problems. In 1899, Hilbert provided an update to *Elements* by Euclid, in his very famous *Grunlage der Geometrie* (*The Foundations of Geometry*), a fundamental, methodological work for understanding modern science. See also Gonseth, F. (1945). *La géométrie et le problème de l'espace*, in six volumes, unfortunately almost impossible to find.

onwards, until its apotheosis, which was the experimental discovery of the Higgs boson at CERN, also called the boson BEH, foreshadowed by the theory developed by the physicists Brout, Englert and Higgs. But the order of magnitude of the particles that make up ordinary matter does not increase significantly. In Figure 5.1, we mention an estimated number of atoms in the visible universe, essentially made of hydrogen, on the order of 10^{80}, a figure established by an astrophysicist of the Collège de France, Christian Magnan[11]. I will not discuss his hypotheses here, because whether it is 10^{100} or 10^{1000} will not change anything in the reasoning that follows.

The immense numbers that interest us here can be qualified as the first kind of immense numbers from those we have just discussed; the ones related to information can be qualified as immense numbers of the second kind.

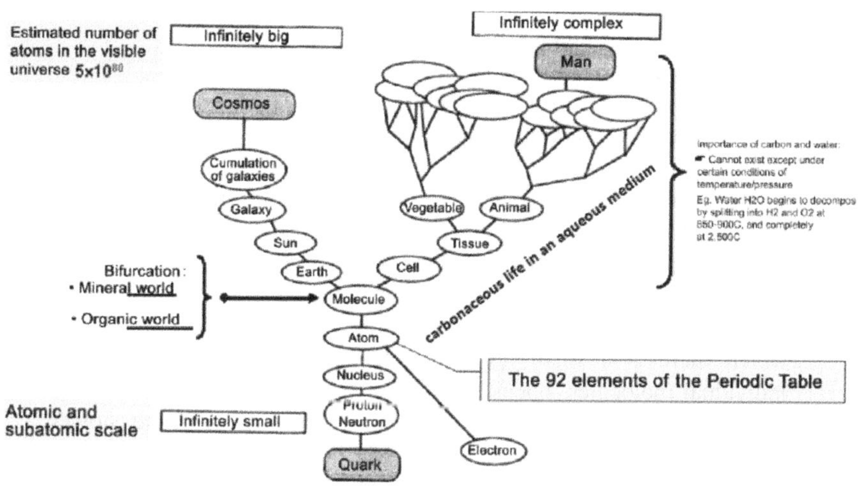

Figure 5.1. *An illustration of the three infinities (source: Gilles Cohen-Tannoudji)*

There is, between these two immensities, something analogous to what so mislead Greek wise men and philosophers, beginning with Pythagoras, namely the rational numbers that we can write in the form of a ratio between two whole numbers, or $\frac{a}{b}$, and those like the number that represents the diagonal of a square

11 See his website: http://www.lacosmo.com/index.html and numerous references on the Web. The origin of this calculation goes back to the 1930s with the astrophysicist Arthur Eddington, who gave a first estimate of 1079. This was also used by Max Tegmark, professor at MIT, in his book: *Our Mathematical Universe: My Quest for the Ultimate Nature of Reality*.

whose side is 1, or $\sqrt{2}$; it is easy to show that we cannot represent this in the form of a ratio even though we may be certain this number exists. The Greeks qualified these numbers as "irrational" since they were contrary to reason, that is, unimaginable in terms of ratios. A problem of the same nature, with the hierarchy of certain infinite sets, was elucidated by Georg Cantor in the 19th century with simple, countable sets, $\aleph_0$, of whole, rational numbers, and the geometric infinity, the continuum $\aleph_1$ that Cantor demonstrated with his famous "diagonal argument" that one is not reducible to the other. The number $\sqrt{2}$ is representative of this geometric infinity that does not run out in the countable[12].

The first caution came from Maxwell himself with his little imaginary being, nicknamed Maxwell's demon, trying to bring some order to the mess of molecules, a thought experiment that challenged the second principle of thermodynamics.

Maxwell, in his thesis on thermodynamics, told us:

> [...] if we conceive of a being whose faculties are so sharpened that they can follow every molecule in its course, such a being, whose attributes are as essentially finite as our own, would be able to do what is impossible to us. For we have seen that molecules in a vessel full of air at uniform temperature are moving with velocities by no means uniform, though the mean velocity of any great number of them, arbitrarily selected, is almost exactly uniform. Now let us suppose that such a vessel is divided into two portions, *A* and *B*, by a division in which there is a small hole, and that a being, who can see the individual molecules, opens and closes this hole, so as to allow only the swifter molecules to pass from *A* to *B*, and only the slower molecules to pass from *B* to *A*. He will thus, without expenditure of work, raise the temperature of *B* and lower that of *A*, in contradiction to the second law of thermodynamics. [...] In questions related to matter, taken on the whole, since we cannot discern each individual molecule, we are forced to rely on what I described as the statistical method of calculation, and abandon the truly exact method, which consists of calculating the movement of each individual molecule[13].

12 See: https://en.wikipedia.org/wiki/Cantor%27s_diagonal_argument.

13 See the reprint by Gabay, called *La chaleur*, based on a translation from 1891, page 421. There is an abundance of literature on Maxwell's Demon; see Leff, H. and Rex, A. (2003). *Maxwell's Demon*. IOP. The first part of the citation was taken from: https://en.wikipedia.org/wiki/Maxwell%27s_demon.

This is what Figure 5.2 illustrates.

Figure 5.2. *Maxwell's demon and the infinitely complex*

In nature, we will never spontaneously observe, on our scale, a situation like the one described by Maxwell. We can, however, using an elementary probability calculation, try to represent the necessary set of conditions so that by the simple product of chance, rapidly moving particles go from compartment A to B. For B to heat up, a significant part of the warm particles from A must pass to B without being compensated for by an inverse movement from B to A, nor can cold particles from B pass to A, thus accelerating the heating of B and the cooling of A.

Here, we have a composite event that is a succession of elementary events, each independent of the others, concerning N particles from A whose speed is superior to the average speed of particles in A and B, somewhat like in heads or tails, with tails representing a particle that has the required properties.

With $N = 64$, we have a composite event whose probability would be written as $\left(\frac{1}{2^{64}}\right)$, since 2^{64} is equal to 18,446,744,073,709,551616 (18 billion billion), and in the notation of engineers, it would be something like $1.8\ldots \times 10^{19}$. An infinitesimal probability, with only 64 particles to pick from, is nothing compared to Avogadro's number N_A, with an entirely undetectable rise in temperature. If we have, for example, 1 liter of water (1,000 grams), on the one hand, and we observe the corresponding event of 18 grams passing from A to B, we are considering an event that corresponds to $2^{6.02...\times10^{23}}$, this time with an absolutely colossal number, representative of what we qualified as an immense number of the second kind. In the notation of engineers, this number would be written as $\approx 10^{180,000,000,000,000,000,000,000}$, that is, of an order of magnitude of 180 thousand billion billion, to be compared with our estimated number of atoms in the universe, 10^{80}.

To become familiar with these notions, beyond the game of flipping a coin, throwing dice or playing bridge, we can use the Galton board, a device invented by Francis Galton (1822–1911, a cousin of Darwin), which is more pedagogical, and for which good animations can be found on the Web[14].

The bead that falls in the triangle, as indicated in Figure 5.3, encounters obstacles in the form of cones set up in the shape of a triangle, such that as the bead hits one cone, it will continue randomly towards either the right or the left, and this will repeat at each of the levels. Since the bead has no memory of the path it has taken nor of the shocks it has already encountered, the choice of left/right made at any particular cone is a "random" choice without finality, and for which the initial conditions are systematically identical; this is what we call "pure" chance.

When it arrives at the 13th level in the figure, the bead falls into one of the 13 columns. When we toss in a bead, we can be certain it will reappear in one of the columns, but we do not know which; the path taken is an image, a symbol, to imagine what we commonly mean by "chance". We should note that each path taken can be represented by a binary number of 13 digits, interpreting 0 as "go left" and 1 as "go right". There are therefore 2^{13} = 8,192 possible paths in the triangle in Figure 5.3; the leftmost path will be written as 0000000000000 and the rightmost as 1111111111111, with all the intermediary possibilities over the course of different tosses.

If we watch the animations, we can see that the filling of the columns is itself random, that is, there is no regularity, but if we make several tosses, there is a convergence towards what is called a binomial distribution, that of the coefficients of Pascal's triangle, which we can superpose onto Galton's board. This is what we call the law of large numbers, a phenomenological law that is never contradicted, except when the underlying phenomenon is not random.

14 See: https://en.wikipedia.org/wiki/Galton_board.

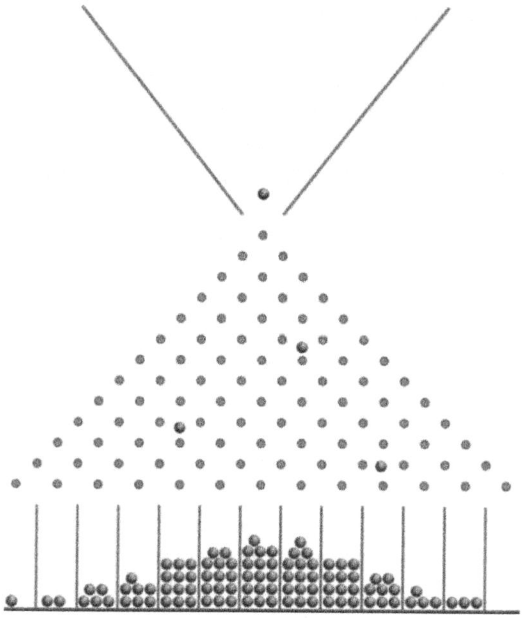

Figure 5.3. *The Galton board*

REMARK. 5.1.– Pascal's triangle corresponding to the figure would be that of the binomial $(1+1)^{13} = 2^{13}$ as indicated previously, that is:

{1, 12, 66, 220, 495, 792, 924, 792, ..., 12, 1}

The probability that the bead will exit through columns 1 or 13 is therefore 924 times smaller than that of exiting at the center from column number 7. A column includes all the combinations of $\{0,1\}$ corresponding to the regular expression $a^p b^q$ with $p+q=n$ (here, $n=13$)[15].

If we smooth out the discrete form that the triangle lets us visualize, we find the so-called bell curve, or the Gaussian curve, named after Gauss, who found its equation[16]. This curve is a practical estimate for calculations, but it is not the reality (see the discussion on Bézier curves, below).

15 Simplified grammar used in computing; see: https://en.wikipedia.org/wiki/Regular_expression.

16 See: https://en.wikipedia.org/wiki/Binomial_distribution and https://en.wikipedia.org/wiki/Normal_distribution.

5.2.1. *Quantum interlude: statistical physics*

The Galton board, indirectly, lets us simulate the interactions that occur in the atomic sphere[17]. There is a relationship between the size and the form of the beads (perfectly spherical, ellipsoidal, irregular, etc.), the size of the cones and the distance between the cones. If the beads are small, the percentage of beads deviating to the right or left will be small, or even close to zero, and they will fall vertically with an occasional deviation to the right or left. Conversely, if they are too big, they will bounce as they would against a wall. The intermediary situation is the most interesting. If the beads are the correct size, but slightly smaller than the holes in the grid, they will go to the right or the left, randomly, and be considerably slowed down, but we are sure they will come out given the principle of conservation. These are exactly the cases Borel described in his book, *Le hasard*[18], where the causes of deviations are so numerous and varied that they disappear into a communal phenomenon that we can qualify as "random", creating the dispersion we observe.

Depending on the scale of observation, one specific property of the observer, the phenomenon can take on the appearance of a continuum as with the persistence of images in the retina[19]. Below the threshold for perception, the phenomenon disappears from our sphere of control because it is not something specific for the observer. The latter may be sensitive to the accumulation of individually imperceptible stimuli, however, whose sum – which requires the ability to memorize – ends up passing the threshold of perception (see Feynman's text at the end of the chapter).

With atomic theory and even more so with quantum mechanics, we know that *observing* in fact means *perturbing*, like poking the hornets' nest. We can understand that the perturbation created by observation may reproduce a macroscopic image of the interaction, which no longer has anything to do with reality since to go from the nano-sphere of individual, minuscule bodies, we need an amplification that will cause billions of billions of intermediary bodies to sway, to create a phenomenon that is equivalent to our scale, such as what physicists accomplished with the Wilson cloud chamber, then the bubble chambers, and finally, the multiwire chambers imagined by Georges Charpak that won him the Nobel Prize in 1992. These devices are at the heart of the ATLAS and ALICE

17 See: https://en.wikipedia.org/wiki/Geiger%E2%80%93Marsden_experiments; a fundamental experiment that showed that matter, in the common sense, is made of "empty space".

18 Published with Félix Alcan, 1st edition in 1914; mine is from 1932, well worn, but now everything can be downloaded...

19 See: https://en.wikipedia.org/wiki/Persistence_of_vision.

detectors in the LHC at CERN[20]. Without these mechanisms, the nano-sphere would have remained a mystery to us!

The theoretical debate that takes shape behind these common-sense considerations is over how to articulate the discrete, *countable* sphere – due to quantification and the specificity of the observed phenomena and the observers (each with their own scale and their own reference framework, even if in a complementary relationship) – and the spatiotemporal sphere, the *geometric continuum* that we are familiar with through our senses. These two worlds are incommensurate, in the literal sense of the term, something which as we know left the Greeks perplexed[21]. We know, since Cantor, that the sphere of the geometric continuum is infinitely more populated, or richer – but these are only metaphors which we must take with a pinch of salt – than the countable sphere, which we represent with whole numbers, however large! This is not without consequence for what we are concerned with in the present work, because a text, like those of the library of Babel, below, is also a number; a pixelated image is also a number. Rational numbers, those that we can represent with a fraction $\frac{a}{b}$ as the Greeks taught us, have the property of being periodic numbers if we represent them using a decimal notation, which can provide good approximations as for the number π[22], for example, which can be written as the fractions $\frac{22}{7}$ (= 3.14*285714285*... with a decimal (followed by 142,857)) or $\frac{355}{113}$ (= 3.141592*920*... or seven significant figures and a period of at most 112 figures, therefore a fairly good approximation).

The problem, clearly explained by Borel in his book *Le hasard*, is that if we make random selections with an alphabet {0,1,2,3,4,5,6,7,8,9}, the probability of picking rational numbers, representable by fractions, tends towards 0[23]. The moral of the story is that random selections, another name for mathematical chance, will always hide rational numbers from us – those numbers which are of our world and our computers.

---

20 These are excellent examples of what Feynman calls "constructions" in the text in the appendix.

21 See the article by Ferdinand Gonseth, "Remarque sur l'idée de complémentarité," *Dialectica*, 1948.

22 See, in the essay collection by F. Le Lionnais, *Les grands courants de la pensée mathématique*, the chapter written by Paul Dubreuil, "L'histoire des nombres mystérieux"; π and some others.

23 Cited above; found in Chapter VII, "Les sciences mathématiques."

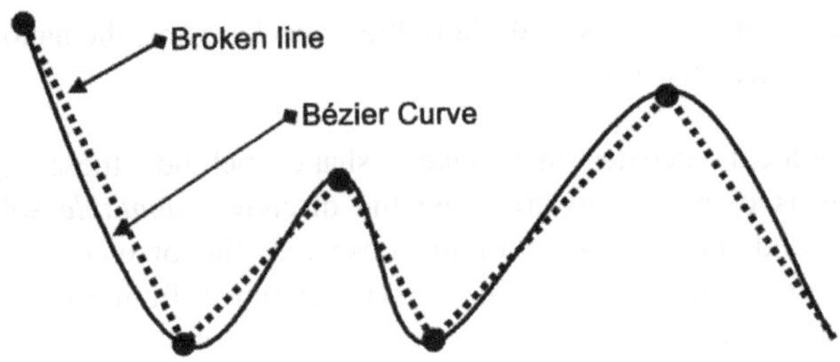

Figure 5.4. *A Bézier curve*

Now, the continuum is, mathematically speaking, more interesting since we have clever mathematical operations for getting there, as, for example, with the Bézier curves[24], which make it possible to go from the discrete – the broken line in the figure – to a continuum, whose use is omnipresent in all the graphic interfaces on our computers today[25].

5.3. Familiar numbers

We are constantly encountering inaccessible numbers without being aware of it, but these numbers have intrigued us for a long time. The number 2^{64} was encountered by the Brahmin Sissa when he tried to be paid in grains of wheat placed on the 64 squares of his chessboard[26]. In China and Japan, the game Go, very popular, takes place on a table, the *goban*, which has a grid of 19 horizontal lines and 19 vertical lines drawn on it, denoted 19 × 19, which create 361 intersections[27]. They say, but I have not verified, that there are approximately 10^{120} possible games in chess – an estimate made by Shannon – and 10^{170} in the game of Go, which makes us consider the latter as "more complex" than chess, even without defining the metrics.

24 See: https://en.wikipedia.org/wiki/B%C3%A9zier_curve; my colleague at CNAM, Professor François Dubois, has told me that for these surprising curves, the radius of the curve of the osculating circle must continuously vary to be able to adapt to the broken line, curvature and torsion in 3D.

25 See, in the collection *Leçons inaugurales* from the Collège de France, the lesson by Jean-Daniel Boissonat, "Géométrie algorithmique : des données géométriques à la géométrie des données," 2017.

26 See: https://en.wikipedia.org/wiki/History_of_chess.

27 See: https://en.wikipedia.org/wiki/Go_(game).

In order to proceed correctly and sort through things, Maxwell's demon needs information about the behavior of particles; likewise when playing chess or Go, we must not only know the rules of the game, very simple in the case of Go, but above all, we must have lots of practice and memorize the greatest number of possible games, that is, sequences of moves against the adversary that almost certainly lead to a victory.

To acquire some familiarity with immense numbers, some formulas from combinatorics and group theory are necessary.

The golden age of combinatorics was Pascal's era, when the theory of probabilities was born, called the theory of chance or of good fortune since it was used to calculate winnings in games of backgammon, roulette, die, etc[28]. For that, the number of players allowed by the rules of the game must be counted, and the winnings can be calculated from there.

Pascal's triangle is one of these well-known instruments that is part of the curriculum during the final years of high school in France[29]. With it, we can easily calculate all the combinations of N objects taken individually, one by one, then two by two, etc., until we reach the end, taking them all together. The corresponding numbers are the coefficients of the Pascaline binomial arithmetic, or $(a+b)^N$ when seen with Galton's board, above. By making a and b equal to 1, we arrive at a formula that is well-known in group theory that gives the number of combinations that we can make with a group composed of N elements, that is, 2^N or $2^N - 1$, if we remove the vacant combination. Since N is the exponent, the number of elements in all of the parts will grow exponentially, as with the functions that carry the same name, that is, a^x, also called a geometric progression. The sorting of particles done by Maxwell's demon shows us that we very quickly arrive at immense numbers.

Another one of these numbers that appears constantly as soon as we sort and categorize, a very frequent activity in all cultures, is the number of possible permutations of N objects, as, for example, in the number of possible hands for a game with 52 cards. However the cards are dealt, one by one, two by two, three by three as in bridge, or however else, we will always be in a combinatorial situation produced by the possible number of permutations for 52 cards. This number is called a factorial, and it is written, in its abbreviated form, as 52!, that is, the product of $1 \times 2 \times 3 \times ... \times 50 \times 51 \times 52$, numbers that we also find in Pascal's triangle.

28 See: https://en.wikipedia.org/wiki/Blaise_Pascal.

29 See: https://en.wikipedia.org/wiki/Pascal%27s_triangle.

REMARK 5.2.– The function $N!$ is often given by its recursive definition, which is written as as $N! = N \times (N - 1)!$, in other words, a function defined in terms of itself, which is shorthand, but introduces an infinite count into the definition, as in the series that carry the same name, for example, the Fibonacci sequence, $u_{n+2} = u_{n+1} + u_n$ with $u_1 = u_2 = 1$, which makes it possible to study the growth of a population of rabbits[30].

When N is large, the calculation quickly becomes fastidious, but we have a good formula for approximating the solution, developed by James Stirling, a Scottish mathematician (1692-1770): $N! \cong \left(\frac{N}{e}\right)^N \sqrt{2\pi N}$ [31]. It is used in pocket calculators, which means that 52! = 8,065… × 10^{67}; that is huge if we wish to run through the combinatorics of playing bridge with games that can last 15–20 minutes, for a total of ≈ 10^{73} seconds! If we wish to run through the combinatoric, compared to the age of the universe, and the number of players in bridge, it is not a problem… plus, there are also the inevitable repetitions of drawing random cards.

Taking a watch with complications such as the Patek Philippe Caliber 89[32], which contains 1,728 parts, a calculator is no longer useful since the number of possible permutations of its 1,728 parts, if we plan to try to put it back together randomly through trial and error, goes beyond its ability to calculate, even using the notation of engineers, and we must calculate it by hand using the Stirling formula, that is: $1.728! \cong \left(\frac{1.728}{2.718}\right)^{1.728} \sqrt{2 \times 3.14 \times 1.728}$ (in other words, any calculation gives $\cong 1.07 \times 10^{4.646}$, if I am not mistaken!).

These numbers are, in the proper sense, absolutely unimaginable and to imagine them, we must have a point of comparison, a reference scale. If we take, for example, the age of the universe expressed in seconds, the 13.7 billion years estimated by astrophysicists gives approximately 4.32… × 10^{17} seconds[33]. If we try to do this by the second, the universe is not old enough to try to put our watch with complications back together. And if it only contained 52 parts, like our card game, it would not be possible either!

---

30 See: https://en.wikipedia.org/wiki/Fibonacci_number.

31 See: https://en.wikipedia.org/wiki/Stirling%27s_approximation.

32 See: https://en.wikipedia.org/wiki/Patek_Philippe_Calibre_89.

33 This duration is compared to that of elementary particles such as the proton and the electron, estimated to be 1036 sec or 1034 sec (basically infinity!) according to the authors, while that of a neutron is only 15 sec.

We must try another time scale, and this time we are going to try it with the Planck constants[34]. These numbers are interesting because they tell us the best that we can expect in terms of interactions given the quantum mechanics that govern our energetic universe at the microscopic scale. We thus have a Planck time of 5.391×10^{-44} seconds, which is the smallest quantum of time that we can hope to get according to quantum mechanics. In Planck units, the age of the universe becomes approximately 8×10^{60} UTP; at this scale, we have some leeway with a watch containing 52 parts, but there is no hope for the Patek Philippe!

REMARK 5.3.– As a reminder, the Higgs boson has an average lifetime of 10^{-24} seconds; it is unobservable with the LHC's detectors, but the consequences of its disintegration are!

In order to have a sense of these numbers which defy imagination, as Boltzmann would say, some combinatorial formulas are useful.

The Stirling formula, whose application we just discussed, is one of the most important.

Another useful formula is the one that gives the average number of draws from a dispenser containing N types of balls for the first repeat to appear, $d = \sqrt{\pi \times \frac{N}{2}}$, and for which we will provide an immediate application with the birthdate paradox, a classic probability calculation exercise whose result is counter-intuitive for students[35]. In a class of 40, students are asked to calculate the probability of two students being born on the same day, which is equivalent to drawing from a dispenser containing 365 types of balls, since there are 365 days in the year. The calculation gives 24, that is, at the 24th draw, there is already a good probability of getting our two students, and at the 40th, it is almost certain.

We have already seen the formula that gives the cardinal of the set of parts of a set of N elements, or 2^N.

We can also provide the formula that specialists of classification are very familiar with when they must find the criteria for arranging N objects (they also say individuals) into k boxes, a common operation for data engineers who develop databases. The number of possible arrangements is $r = \frac{k^N}{k!}$, which combines parts

34 See: https://en.wikipedia.org/wiki/Planck_units.

35 This kind of formula is called asymptotic because these are finite numbers, as with the approximation of $N!$. Its demonstration is generally rather difficult.

and permutations together. We can verify that the number r also grows very quickly. Bouroche and Saporta gave the example of the classifications that are possible for 14 individuals who must be divided into four classes, which gives $\frac{4^{14}}{24} \cong 11.184.811$ possible groupings. Since a classification always has an element of arbitrariness in the choice of criteria, two classifiers left to their own devices will produce two different classifications. The consensus, or the overview of the set of parts, is therefore a prerequisite to the use of classification by the relevant populations, for without a consensus, there is no reason to be in agreement given the number of possibilities; incidentally, we can see that this is a real challenge for democracies which operate on the basis of a consensus accepted by all.

Finally, we must keep the formula in mind that indicates how reliable a series of connected processes are, also called the theory of the "weakest link" among the quality control and assessment community. Each of these processes has a reliability of $r < 1$, which means that the reliability of the chain of processes is $R = r_1 \times r_2 \times \ldots \times r_N$, from which we can deduce that with $R < r_x$, r_x being the "weakest link" in the chain, and moreover, with all r being < 1, R unavoidably tends towards 0[36]. With computers, which execute thousands of transactions (processes) per second, the occurrence of an error is therefore inevitable, hence the mechanisms for repetition, integrated in operating systems since the beginning of the 1970s. For the error to appear, it is very simple: all you have to do is wait, and since computers go faster and faster, and work more and more in parallel, the waiting time for an error to appear is shorter and shorter, but we know how to compensate for these random occurrences with "intelligence" when we go about it in the correct way. In critically important systems, there are what is known as "majority vote" mechanisms, founded on the random nature of certain errors, whose principles will be examined in section 6.2.1.

5.4. The library of Babel

Jorge Luis Borges is the inventor of a "total" library, described in his short story as "The Library of Babel"[37]. All of the books in the library have the same format, 410 pages, 40 lines per page and 80 characters per line. The typography is reduced to 25 characters, including punctuation. Anecdotally, a book of 410 pages is almost the size of one of my recent books, *Architecture logicielle*, published with Dunod, that I spent almost three years writing; it contains 494 pages in its 3rd edition, with a richer typography and many figures.

---

36 See: https://en.wikipedia.org/wiki/Six_Sigma.

37 See the collection, *Fictions*, published by Penguin Classics (2000, p. 65).

The library of Babel contains every book possible, all in the prescribed format, corresponding to a combinatorial written as $25^{1,312,000}$, with each volume containing $410 \times 40 \times 80 = 1,312,000$ characters. Such numbers are unthinkable in the literal sense, but by trying to imagine them, we can calculate their order of magnitude, for the very large ones, in the engineering notation using a formula known by high school students thanks to logarithms: $a^N = 10^{N \times \log(a)}$; the integer part of $N \times \log(a)$ gives the number 0 as the exponent, and the decimal part $10^{0,xxxx...}$ gives us a number from 1 to 10 that is essentially negligible.

The number of books in the "total" library of Babel is therefore, in this notation, approximately $\approx 2 \times 10^{1,834,097}$ volumes!

If we look at the material side of things, we would need many writers to write all the books, even if some of them could be produced automatically by computers, or even by monkeys, to use Emile Borel's metaphor in his treatise on probabilities called *Le hasard*[38]. In order to represent them in the physical reality of our universe, we need at least one atom in several states, or several atoms whose states are distinct (ground/excited) to get our 25 characters, or five atoms ($2^5 = 32$). We return to the pitfall that we already encountered: there are not enough atoms in our universe to even represent an infinitesimal part of the library of Babel.

Borel imagined an army of 1 million monkeys randomly typing on a million typewriters and asked himself if it would be possible to calculate a probability of coming up with some kind of literary or scientific work, and how much time it would take. Here again, we surpass the acceptable limits of our universe by far; but again, we would have to account for the phenomenon of repetitions cited above with the birthdate paradox. In the immense space that could contain all the types of books, if we were to select one randomly, it is the formula for the repeat that we have to use, $d = \sqrt{\pi \times \frac{N}{2}}$, with N being the number of books; the notation $\sqrt[2]{\ }$ is equivalent to dividing 1,834,097 by 2, but that gives even more than 917,000 zeros, so it is still off limits, except that this time, we have produced a book that already exists in the library, but without knowing which one!

38 See: https://en.wikipedia.org/wiki/Infinite_monkey_theorem; in some of his other works, Borel speaks of "imagined numbers" or even "inaccessible numbers". See his books *Les nombres inaccessibles* and *L'imaginaire et le réel en mathématiques et en physique*. The book on chance, *Le hasard*, is almost impossible to find, but can be downloaded thanks to the "magic" of the Internet.

Even working at the speed of interaction defined by Planck units, we still cannot reach the count, and by far, even with our most powerful computers, such as the ones used to sort the data produced by the LHC at CERN.

We can say that the numbers of the first kind are the ones that we can represent materially, and manipulate, using the resources that our physical universe offers us – let us say at the Avogadro scale – recalling that we are limited by Planck units and the size of the universe. All of the numbers represented in our computers are of this type, by definition. The numbers of the second kind are all the others, in particular those that we find in information sciences. We can realize these numbers thanks to mathematics, and we can even give them a name, but they will never have a concrete, physical existence; they are "supernatural", in a sense! In homage to Borel, we will use his terminology of *inaccessible numbers* to designate them from this point forward.

5.5. Inaccessible numbers of the third infinity: information sciences

At the time when Emile Borel was imagining his typist monkeys, there was no talk of a third infinity, but people already suspected that living organisms were much more complex than what may have been imagined up to that point, without really knowing how or being able to quantify them[39]. One of the first authors to speak of the third infinity in the sense presented in Figure 5.1 was Teilhard de Chardin, in his book *The Phenomenon of Man*, now translated by Julian Huxley, but which was never published during his lifetime. During the 1950s and 1960s, the matter became clearer thanks to two major discoveries: computers, on the one hand, and the genetic code with its alphabet of four letters, made of the bases A, T, C and G or simply ATCG, on the other. The science of the living, ontogeny, which certain biologists are beginning to approach as such, remains a mystery, but we know a part of its foundation, DNA molecules of chromosomes and mitochondria[40]. In many ways, this science remains a "black box" whose efficiency we can nonetheless observe in ontogeny. Genomics of the 1990s owes everything to high-performance computing, at our scale, which emerged in those years! Conversely, complex artificial systems engineering, of which one of the most successful products, the smartphone, is accessible to everyone nowadays, is an example of perfectly controlled engineering, every aspect of which has been mastered, and which integrates the fundamentals of the knowledge and know-how of the 20th century.

39 See p. 164 of his book *Le hasard*, Ed. Alcan, 1932.

40 See, for example, the book by Philippe Kourilsky (2014) and in another register, Jean-Louis Dessalles et al. (2016), *Le fil de la vie*, Odile Jacob.

The two groups of people concerned by these numbers are, on the one hand, the software engineers who must live and cohabit with them, since they constantly make things with them while trying to control them; and, on the other hand, biologists, who must deal with them or observe them ever since the discovery of the genetic code, and then try to give an understandable and useful image of them using the software. We will give some examples for each of these groups.

5.5.1. *The numbers of software engineers*

The principal organ of all modern computers is the memory, which makes it possible to store information so it can be used, that is, to make calculations, in the wider sense of the term (this is the Church–Turing thesis), using programs that are themselves saved in the memory[41].

The size of available memory has grown considerably, more or less at the rate of Moore's law, that is, it has doubled every two years over 40 years, or 2^{20} (or approximately 1 million times). But in the 1950s–1960s, let us say before the arrival of micro-electronics and VLSI, the combinatorial phenomenon was already part of the everyday life of programmers[42]. Incidentally, the computer on which I wrote my first programs existed in three versions: 2,500, 5,000 and 10,000 octets of memory (made with ferrite cores since at the time, transistors were too big), with decimal addressing visible to the programmer via what was called an assembly language, only available on this computer at the end of the 1960s. The "biggest" computers at the time did not surpass 32,768 octets (or 32K, as we would say) with binary addressing of 15 bits ($2^{15} = 32 \times 1{,}024 = 32K$).

In one octet, 256 elementary pieces of information can be coded, which means that a computer with 5,000 octets has a coding capacity of $256^{5,000}$, or in engineering notation, applying the formula from above: $10^{5,000\times\log(256)} = 10^{12,041.1998266}$ $= 1.584\ldots\times 10^{12,041}$, if we believe my calculator[43].

We are now encountering immense numbers of the second kind, that is, inaccessible numbers.

---

41 For an introduction, see the short book by Gilles Dowek, *Les métamorphoses du calcul*, cited above.

42 *Very Large Scale Integration*. See the book by my colleague, François Anceau, *Conception des circuits VLSI*, Dunod, 2007.

43 See: https://en.wikipedia.org/wiki/Octet_(computing).

Today, among the technology that is available, a memory of 32 gigabytes (GB), or 32 billion octets, is something common. Let us now do the same calculation with 1 GB of memory instead, or 1,073,741,824 octets ($(1K)^3$, or even 1,024 × 1,024 × 1,024). The number of possible states of this memory is therefore $256^{1,073,741,824}$, or in engineering notation, $9.55... \times 10^{2,585,827,972}$; we can certainly say that things start to fall apart! From the start, the programmer finds themselves in a "library of Babel" type of situation, even if they do not realize it.

What is interesting about programmers is that many of them think that with the processing power of the computer, say, 1 billion commands per second, we will always be able to "cope with" things, the way Rube Goldberg would have done[44]. If we want to explore this combinatorial system with a 1 gigahertz computer, which is a measure of the speed per second, all we have to do is divide by 10^9, and we get the number of seconds. If we have 1 billion computers of this power, we redivide by 10^9 or 10^{18}, and we unfortunately conclude that the universe is still not old enough, and by far.

The moral of the story is that no properly trained programmer with a good understanding of the unique and new side of their job will work in this way, and if they are not rigorous, they will immediately fall into the trap of combinatorics. This is a field that emerged in the 1950s, during which programmers could be counted by the dozen, whereas today they number in the millions, and whose key element is the architecture, the only way of organizing complexity. A large part of the issues blamed on programming comes from the lack of understanding of this phenomenological law, as well as from the inveterate optimism of many programmers who think they can "cope" through repeated attempts that will still be infinitesimal in number with respect to the combinatorial[45]. Complexity must be organized through its design (see Chapter 1).

We will see how, with the metaphor of the computer stack and thanks to the methods developed along the way, programmers created a tool for making "infinitely" complex objects such as smartphones, the methods being founded on the general principles from cybernetics and the constructive logic of the theory of automata, integrating the "more is different" principle of the physicist Anderson (1972). The magical formula for avoiding risks when working with a combinatorial of inaccessible numbers, this third infinity that surpasses the capacity of physical nature – in this sense, they are "supernatural", that is, beyond what

44 See Rube Goldberg's "machines" online: https://en.wikipedia.org/wiki/Rube_Goldberg_machine.

45 For more details, see my book *Ecosystème des projets informatiques*, Hermès-Lavoisier.

nature can show us – is none other than organization, architecture, creativity, patience (not wanting to skip steps, proceeding step by step), adaptability, not over-thinking or over-complicating methods (this is where Occam and his razor, or his alter ego, Wittgenstein, reign supreme), that is, in a word: intelligence. A programmer who has truly understood the nature of the abstract entities that they are creating reasons in terms of functions and composition of functions, carefully distinguishing abstract expression from concrete materialization through a translation/transduction interface.

A few words on a class of new entities that appeared concomitantly with the first computers, that is, the programs written by programmers, in the programing sense of the term, as summarized by Niklaus Wirth, the inventor of the PASCAL language:

programs = data + algorithms

In this slightly lapidary formula, an algorithm is a series of commands (in the jargon, we say *statements* or even *instructions*, a term which we find when we speak of instruction set architecture) that apply to certain data provided at the start of the calculation process, which will arrive at a result that will be made up of other data at the end of the process[46]. Hence, the alternative notation:

$$Programs \Rightarrow E(e1, e2, \ldots, e_n) \xrightarrow{Algorithms} S(s1, s2, s3, \ldots, s_m)$$

An algorithm can itself have internal data specific to it, for example, the equivalent of a logarithm table such as those used by the engineers working before the era of pocket calculators.

REMARK 5.4.– The algorithm that calculates taxes in western countries uses the tax brackets defined by the tax code and the fiscal law voted on by representatives as its internal data.

No program can exist without a computer that is capable of interpreting the program, or in other words, executing the program. It is the computer that determines the meaning of the program; and inversely, the text baptized as a "program" without a corresponding computer is nothing but incomprehensible

46 In the 1980s, computer designers put together reduced instruction set computers, RISC, as opposed to computers with complex instruction set architectures, or CISC. With the abundance of transistors, this debate began to die out in the 1990s and has completely disappeared today. It is common to have computers with instruction sets of several hundred instructions.

gibberish. Nevertheless, if we are sure that the text in question is a program, then we can seek a computer capable of interpreting it, which is what Alan Turing did in decrypting the messages of the German army's Enigma machine during World War II, or what Champollion did in decrypting the Egyptian hieroglyphs with the help of the Rosetta stone and his knowledge of Coptic[47]. Hence the rule of reverse engineering established by Turing for encryption: "what a computer does, another could undo" (as a reminder, a telex page of 4,000 characters is a combinatorial of $50^{4,000}$ for 50 typographical signs, or $10^{6,796}$, in other words, an immense number).

Data and algorithms are aspects, or the dimensions, of the entities known as "programs" that make up a whole; they are indissociable even if distinct, and in a "complementary" relationship, in the quantum mechanical sense of the term[48].

To write a program is to simultaneously define the data and algorithm that will create the desired correspondence $E \xrightarrow{Algorithm} S$, given that the data determines the algorithm, and vice versa.

From here, we arrive at a new variation on inaccessible numbers, that is, the number of possible programs given the sets *E* and *S*, a number we can obtain using a formula from set theory: $\#P = (\#S)^{\#E}$, where the sign # means "the cardinality of...". #*E* and #*S* being themselves inaccessible numbers, the number of programs *P*, or #*P*, is itself an exponential of inaccessible numbers! In this combinatorial immensity, only a subset of #*P* really corresponds to the wishes of the programmer and their boss, and since programmers, like everything that exists in nature, will make errors, we are never sure of having the correct program that functions in every scenario, including when the data provided is incorrect and/or corrupted. Programs are the finalized objects that respond to a need as defined by a user, that is, a function/action described in a qualitative way using examples and/or flowcharts. The program thus appears to be a generalization built through successive refinements, modulo the basic instructions of the computer, and the functions available in libraries – what we call APIs (Application Programming Interface) in programmer jargon. The validity of a program depends on whether it conforms to an expressed need, hence the systemic cycle of needs ↔ programs that may begin somewhat sloppily, but whose necessity ultimately determines its convergence towards an acceptable solution.

---

47 See the reference bibliography by Andrew Hodges, *Alan Turing: The Enigma*, and the overview of articles and works by Turing, collated by Jack Copeland, *The Essential Turing*, Oxford University Press.

48 See, in particular, the special issue of *Dialectica*, 2(3–4), "L'idée de complémentarité," containing an overview by the brilliant F. Gonseth.

The situation programmers find themselves in, all things otherwise being equal, is much worse than that of their hardware logician colleagues, for whom the world is made of 0s and 1s, since to transform one bit into another bit, there are only four functions, or $2^2 = 4$. To transform two bits (two entries of one bit), they have a choice of 16 functions, or $2^4 = 16$; and with n bits, they have exactly 2^{2^n} functions to choose from. If we decide to try making modal logics with 3 or more valid values, as, for example, in {true, false, indefinite}, the number of possible programs "explodes" to 3^{3^n}; this incidentally shows the intellectual impasse of the theory of fuzzy sets that get called "fuzzy" since they cannot be verified due to the combinatorics they produce. Programming, which has largely developed in a parallel, generalized and asynchronous way, as is found in cloud computing and in new-generation multi-core chip designs, capable of executing thousands of programs in parallel (called *transactions* or *threads* in the jargon), has made the situation all the more precarious (see Figure 6.7).

There are natural solutions to avoid getting lost in the immensity of combinatorics and thus nonsense; this is what creates the scale and the difficulty of a programmer's job, which is everything except simple as we must learn to move forward through the infinitely complex without getting lost, but this is an altogether different story (see Chapter 6).

5.5.2. *The numbers of the genome*

In the engineering of life, there are numerous decisive steps related to the infinitely complex. The theory of evolution and the data from paleontology prove that living creatures are the fruit of an evolution spread over more than 2 billion years. All living beings that exist on Earth have one common ancestor. Pasteur, a chemist by training, demonstrated the importance of biochemistry using cellular machinery, bacteria and eventually viruses, invisible under light optical microscopes. There are approximately 10^{14} cells in a human being, and more or less the same number of bacteria[49]. But the watershed moment was the discovery of DNA and the "genetic code" in the 1950s. Hence the notion of a genetic program that spread, especially through the work of Jacques Monod, *Le hasard et la nécessité*, a notion that has today been questioned, and even called erroneous, by some[50]. We now have a complete map of a large number of genomes[51]. We will examine five cases of this nomenclature from the point of view of combinatorics and its supposed complexity.

---

49 See the book by Pascale Cossart, a member of the Academy of Sciences in France: Cossart, P. (2016). *La nouvelle microbiologie*. Odile Jacob, Paris.

50 For example, Jean-Jacques Kupiec, among others.

51 See: https://en.wikipedia.org/wiki/Genome_size.

The first case: the bacterium *Escherichia coli*, also called *E. coli*, is an intestinal bacterium found in mammals, and very common in humans; it makes up approximately 80% of our aerobic intestinal flora. Its genome contains 4.60 million ATCG letters. One eucaryote cell such as the paramecium contains approximately 72 million base pair letters. These genomes are enormous to begin with, and thinking of them in terms of the scale of the library of Babel, that would bring us to 4 and 55 books respectively!

REMARK 5.5.– Viruses, which are cellular parasites, are much smaller: there are approximately 15,000 base pair letters in the common flu, and 31,000 in Covid-19.

The second case: the *Drosophila melanogaster*, still called a fruit fly, is a Diptera insect (therefore having only one pair of wings); it is the type of drosophila used commonly in genetics experiments to study hereditary traits, beginning with Thomas Morgan at the beginning of the 19th century. This organism is one of the most frequently used models in biology/genetics. Its genome contains 118 million ATCG pairs.

The third and fourth cases: two mammals, the mouse and man, that have the commonality of having almost the same number of base pairs in their genome, 3,400 million (3.4 billion; but according to other sources, 2.7 billion for the mouse). What counts is the magnitude of size, which here, is in the billions.

Finally, the fifth case: the largest known genome is that of the Japanese flower *Paris japonica*; it contains almost 150 million ATCG base pairs, almost 50 times the size of the human genome.

At the scale of books in the Library of Babel, the human genome represents approximately 2,600 books, all of which can incidentally be found somewhere in the library, as with all other genomes.

Between bacteria and unicellular organisms, genomes are in a ratio of 10:20, on the same order of the drosophile's compared to mammals, where we have a ratio of 20:30. This shows that we must be prudent if we wish to establish correlations between size and apparent complexity.

Nevertheless, when we say that human and chimpanzee genomes only differ by 1–2%, we must reason using combinatorics in order to avoid being misled by the percentages. A 1% difference means 34 million ATCG different base pairs, and in combinatorics, that gives us $4^{34,000,000}$ possibilities, which in the notation of engineers is approximately $5 \times 10^{20,470,039}$, therefore an inaccessible number. Using the formula for redundancies, it is clear that there have never been two entirely identical humans in the lineage of hominids, except perhaps homozygous twins. But we can do even better by examining the combinatorics of cells.

My colleague at the Collège de France, Philippe Kourilsky, pointed out to me that in a human being, there are probably no two entirely identical cells from the perspective of their genome, due to replication errors made in cells throughout the lifetime of a human, starting with the unique cellular fusion of an ovum and a spermatozoon, differences that do not prevent the organism from functioning. A human being is made up of something on the order of 10^{13} cells that get renewed approximately once a year at variable rates. This means that throughout the lifetime of a human, estimated at 100 years to simplify things, the organism will have made approximately 10^{15} or 10^{16} cells. Soon, there will be 10 billion human beings on Earth, which means that we can estimate that since the appearance of homo sapiens some 200,000 years ago from a small group of a couple dozen individuals, or somewhere between 8,000 and 10,000 generations, the total number of homo sapiens never having existed must be below 100 billion (10^{11}), recalling that during the first dozens of thousands of years, the population of humans was low, on the order of 20,000–30,000 at the time the Lascaux caves were painted. The number of human cells produced would, according to this hypothesis, be something like 10^{26}, but if it were 10^{50} it would change nothing in the rationale: there have probably never been two human cells that were perfectly identical ever since we began to exist. Such is the phenomenology of these inaccessible numbers.

If we retrace these steps using the entirety of the genome instead, the combinatorial is on the order of $4^{3,400,000,000}$, approximately $3 \times 10^{2,047,003,970}$ in the notation of engineers, which only worsens things. We are all definitely and irredeemably unique, including at the level of our cells, and this will probably be the case until our solar system disappears in a couple billion years.

As for the Japanese flower, *Paris japonica* poses an interesting problem since the size of its genome does not reflect the real complexity of the plant when we compare it to the genome of animals. The *Paris japonica* genome contains numerous redundancies, which means that the "gross" size is not meaningful. The underlying issue is how we measure complexity. An analogy in programming gives a clearer sense of these benchmark issues and the questions it raises. Since the very beginning of computer programming, the complexity of a program was habitually denoted by counting the number of instructions needed to write its program. For example, in computer programming, we could count the average number of instructions written in a "source" language such as COBOL and FORTRAN at the beginning, and today, C++, Java, PYTHON, SQL, etc. We can also count the number of command prompts needed for the compiled program to be executable, that is, the number of instructions generated by the compiler. Finally, we can look at the memory image of the program built in the memory after the executable file was downloaded by the operating system of the computer. This means that for one initial text, describing the *what* of the program, we will end up with three different sizes:

– the program source text;

– the binary compiled text, editable in the assembly language, understandable only by "system" programmers (some percentage of the total population of programmers);

– the executable binary, almost incomprehensible by a human operator, whomever they may be.

The size of the compiled binary will depend on a) the nature of the programming language, which is what we call its "level", that is, the capacity for abstraction, specific to each language, and b) the capacities of the computer itself. The general languages known as object languages, such as Java or PYTHON, currently have the best level of abstraction, which on average translates into a statistical rule in which "1 source instruction always produces an average of 4 to 6 command prompts in the compiled binary"[52]. Incidentally, all programming languages since COBOL have copy/paste editing mechanisms, which make it possible to write a sequence of instructions in one unique point and recopy it as is as many times as needed in the program. These mechanisms are fundamental for the evolution of the program throughout the life of the system in which it is integrated. Using these mechanisms, a program of initial size T can double or triple in size, or even grow by 10-fold once all the copy and pastes have been completed. Incidentally, computer programmers invented many specialized languages to define and control SQL data in the 1980s and 1990s, but with expansion coefficients of two orders of magnitude.

As for the size of the executable binary, it is dependent on two principal mechanisms: a) the mechanism for "editing links" that manages the initial program's connections with common functions offered by the operating system where there can again be duplicates/replications given the modalities of recalling the function (sequentially or non-sequentially and statically or dynamically, in the jargon), and in the binary format of the computer; and b) the mechanism for controlling the chain of computer instructions, a mechanism that is today "hidden" from programmers except for those who need detailed access to the intimate structures of the computer, hence the new duplicates/replications.

Finally, for the sake of completeness, there are the capacities of the computer itself. At the beginning of von Neumann's design, there were intense discussions on whether the command "go to", which allowed for calculation cycles to be programmed, was well-founded or not. Without this instruction, a calculation

52 See Capers Jones, *Programming Languages Table*, available at: http://www.cs.bsu.edu/homepages/dmz/cs697/langtbl.htm.

cycle of N turns requires a replication of the corresponding sequence of code *N times* – this is what gets called "unfolding the cycle" in the jargon of computer designers – which means that the size of the corresponding program is virtually multiplied by N. This does not pose any computer code generation problems; however, the memory becomes blocked by a multitude of almost identical command sequences. This command is therefore necessary, not functionally, but for reasons of optimization of calculation capacities, and more specifically for memory capacity. To give a sense of size, we could show that a program of 1 million lines of source code may need approximately 1 billion transistors if the computer has an ad hoc instruction set architecture such as the command "go to", to which we would have to add what is needed to store the data used by the program.

At the very end, we could count in bits, or even in the number of components (transistors, relay ports, etc.), which would have absolutely no meaning from a semantic point of view.

The moral of the story is that a "gross" size expressed as a pure number, without knowing about the operating computer and the mechanisms for optimizing how the different codes are processed, has absolutely no meaning; it is the computer that fixes the energetic "dimension" of the number in terms of assemblage/integration efforts.

To return to the genome of *Paris japonica*, the only thing that we can affirm, if we stick to the software metaphor, is that the "optimization" mechanisms at work in the evolution of genomes of plants and animals are surely completely different, even if the basic "computer" code, namely the quadruplet ATCG, remains the same, that is, the equivalent of bits. In particular, the control of duplicates/replications is crucial for this economy since without these mechanisms there would never have been what software engineers called "programming in the large" or "mega-programming", a part of all the great infrastructures that are now computerized and in which the number of lines of code written by programmers reach the tens of millions. Let us recall that a well-trained programmer, with a few years of experience, has an average productivity of about 4,000 lines of code per year, or 2.5 lines of code per working hour, which retrospectively illustrates the breadth of the human investment that must be made when entering this competition, even if the term coopetition would be more appropriate because it integrates two fundamental aspects of this new technological territory: competition and cooperation, where the collective human dimension is essential[53].

As for the genome and ontogenesis, the final word is obviously left to biologists.

53 The core of my academic work deals with this issue (see the references).

6

Walking through Complexity

6.1. The example of the computer stack

The stack of computer interfaces in a smartphone is certainly one of the most complex artificial objects/systems ever imagined by humankind, if not the most complex. The numbers, seen from the outside, are impressive.

Figure 6.1(a) represents a personal computer, of the kind used at a workstation in the 1990s; the smartphone in Figure 6.1(b) represents the point which we have reached today. Between the first commercial Whirlwind computers made by IBM for the USAF's SAGE project at the beginning of the 1950s and the computers of the 2000s such as the smartphone, performance has been improved by a factor of 1 million, or six orders of magnitude[1]. This is something that has never before been seen in the history of technology[2].

With engraving techniques reaching 10–20 nanometers, a current chip can contain up to almost 6 billion components, connected by a delicate network of cables that enable the computing functions and provide electrical power at a limited

---

For a color version of all the figures in this chapter, see www.iste.co.uk/printz/pedagogy.zip.

1 See: https://en.wikipedia.org/wiki/Semi-Automatic_Ground_Environment; https://en.wikipedia.org/wiki/Whirlwind_I.

2 We should also point out the excellent work by Gilbert Simondon, which deserves a reread, in particular *Du mode d'existence des objets techniques*, as well as the Pléiade volume *Histoire des techniques* by Bertrand Gille. For artificial systems, *The Sciences of the Artificial* by Simon is canonical.

rate of 1–2 gigahertz, as beyond that the circuit would overheat[3]. This means that a modern chip, also called a System on Chip (SOC), contains hundreds of kilometers of nanometric cables in its silicon substrate, initially made of gold, but since mastering the physics, they are made of copper-coated aluminum.

The silicon used in electronic chips is itself the result of a long process of physicochemical preparation requiring hundreds of operations before arriving at an operational circuit, given that it does not exist in the natural state. Even though silicon is one of the most abundant elements in nature, bars of ultra-pure silicon crystal, approximately 30 cm in diameter, one meter in length (about 160 kg) must first be made, which are then cut into round slices, called wafers in the jargon[4]. If we take circuits that are more or less the size of the one in the figure, approximately 500 mm^2, one wafer is enough to make about 100 chips in just one batch. For complex circuits such as the ones that are needed for smartphones, the manufacturing process requires many hundreds of operations before a circuit that is ready to receive its first programs can be obtained, with each step being validated by ad hoc test computers. In the end, the production process takes from six to nine months, depending on the complexity of the circuit.

In a chip, several logics are superposed without contradicting each other:

– A material logic, at the nanometric scale (at this scale, and in the conditions required for the computer to work properly, everything moves, atoms are diffused, the noise of the physicochemical environment perturbs, etc.). Given their nanometric size, components can be perturbed by the radiation produced by the natural radioactivity of alpha particles, for example. Consequently, the components that make up the parts of the circuit are protected by error correction codes (ECC), a theory devised by Claude Shannon.

– An electrical logic, since we are first and foremost dealing with a gigantic electrical circuit made up of billions of components connected to each other and powered at a constant rate (for chips like the ones in Figure 6.1, there are several clocks per chip, one per processor, all of which must be synchronized; this heats up, and the heat must be evacuated given that silicon is a poor conductor of heat).

– A hardware logic, for functions that are "plugged-in", that is, transducers that produce an outgoing signal in response to an incoming signal, for example, the addition of two numbers (note that to add 2 bits, about 30 transistors are needed).

3 We know how to make clocks of 1,000 gigahertz, but we cannot use them in our computers. Present engraving is today round five nanometers.

4 See: https://en.wikipedia.org/wiki/Zone_melting and the article about the British physicist who discovered it, Bernal (https://en.wikipedia.org/wiki/John_Desmond_Bernal), before it began to be used for industrial purposes at Bell Labs by William Gardner Pfann.

these hardware functions are themselves programmed by software engineers in ad hoc languages such as VHDL[5]. These languages are then translated into the languages that control the engraving machines.

– A software logic for the software that activates the plugged-in hardware in order to complete the operations requested by a user external to the computer, that is, external processes split into ever more specific elementary processes until the ones that are implemented in the hardware functions of the computer are found.

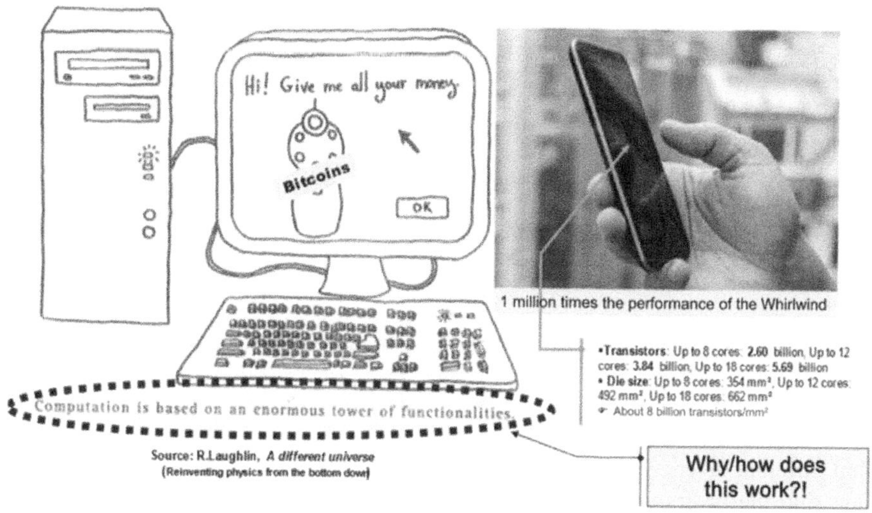

a) b)

Figure 6.1. *From the PC to the smartphone (exterior) (source: Laughlin, R.,* A Different Universe*)*

These different logics are a perfect illustration of the principle of emergence that "more is different", established by the Nobel Laureate, Philip Warren Anderson (1972). Figure 6.2 summarizes the situation described above.

At each stage, we find a library of tests adapted to the specific hazards of the level, something which requires a complete understanding of the purpose of the level, its nominal environment that is being monitored, how it is made and how it is used by the upper levels (i.e. the uses which, by definition, cannot be inferred from lower levels). Each level is dependent on the lower levels via negotiated interfaces (which are "contracts" in the literal sense of a reciprocal engagement) that

5 See: https://en.wikipedia.org/wiki/VHDL.

show us what is functionally necessary in the form of "return codes" that are the "traces" of what happens at each level, at the very least in the form of done/not done, and if not, for what reasons. As we will see, the level programming is itself subdivided and organized into a series of levels, purely in terms of software this time, relating only to the programming, that is, a recursive stack of abstract machines on the basis of the theoretical model initially imagined by Turing.

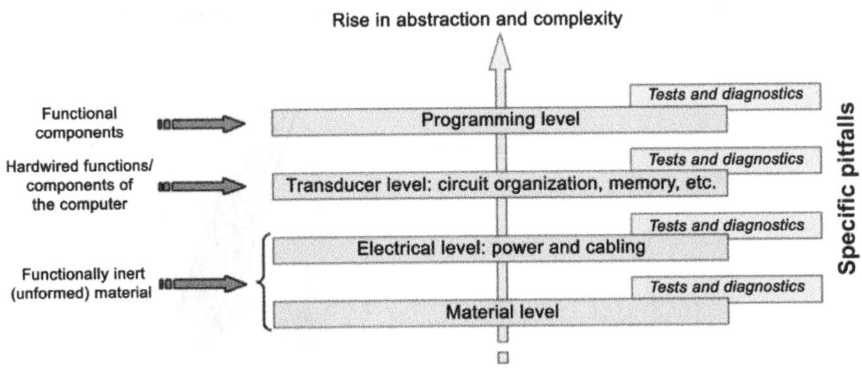

Figure 6.2. *Organizing the cohabitation of different logics*

If this time we look at the software side of things, as indicated in Figure 6.3, we find different layers. At the top of the stack, just above the hardware functions, we first find the layers specifically dedicated to the operating system, and the micro kernels needed by multi-core chips along with the drivers (in the jargon, these are programs called "pilots" because their function is to control the environment) needed by the external inputs/outputs depending on their specificities (graphic screens, voices/images, network access interfaces, editions/publications of results, etc.).

Depending on each computer manufacturer, the operating system and the different pilots needed for creating the interfaces are programs containing millions of lines of source code[6]. For example, the LINUX core, without its pilots, reaches 2 to 3 million lines of code; but that can rise to anywhere between 6 and 8 million, depending on the equipment that is connected.

At the very top of the stack, above the application interface – called API (Application Programming Interfaces) in the jargon – is where the programs created by development teams of the manufacturers responding to their specific needs and

6 The operating system that I worked on for about 20 years contained approximately 4.5 million lines of code, counted according to the method recommended by the SEI.

those of customers can be found. At this level, we can count in the tens of millions, which means that in a basic smartphone, we will find up to 20 or 30 million lines of code written by thousands of programmers.

NOTE 6.1.– The library of applications that a large company has can contain tens of thousands of programs, reaching anywhere between 100 and 200 million lines of code[7].

As with all computers and/or tools invented by humankind, ever since the very first pieces of flint used by our earliest ancestors, computers are finalized objects whose function is determined by the use that is made of them, that is, simultaneously by those who conceive of them and make them and those who use them, the users. In the language of modern engineering, this is what we call co-design. Incidentally, we can see that we are, ourselves, finalized beings, capable of executing a great variety of tasks, resolving problems and creating all sorts of artifacts, adapting to all kinds of environments, from the expanses of the Arctic to the most inhospitable deserts.

The keyword of this grandiose edifice is *design to test*, for a testable conception/architecture, since it is not enough to just build it; it must also "work" or "hold up" and the user must individually or collectively be able to capitalize on it in a real way, with increased efficiency. For that, the designer must create a construction wherein each level can be certified for what it is, independently of the use that will be made of it. At this level of complexity, it is impossible to rely on chance, and all the components must be verified exhaustively, as one error that appears in level N-1 will appear as a bug at level N without the possibility of diagnosing the problem from that level since it will not understand its logic; even more so, when we go through the following levels of abstraction, $N+1$, $N+2$, etc. It is perfectly clear at the level of the hardware/software interface, but it is naturally the case at all levels.

At a nanometric scale, it is impossible to address the problem on or in the circuit manually, as it is too small; its proper functioning must be verified from the outside, which is in essence the same as validating a behavior. One of the most difficult problems that a systems architect must solve is how to create a perfect correspondence between the external, observed behavior using the ad hoc means that they have themselves defined and the internal structure for which they control the entire construction process since they know the physics of its constitutive elements. The quality control process plays a central role in this overall process of integration.

7 See the CESAMES website for the white paper, *Intégration et complexité*.

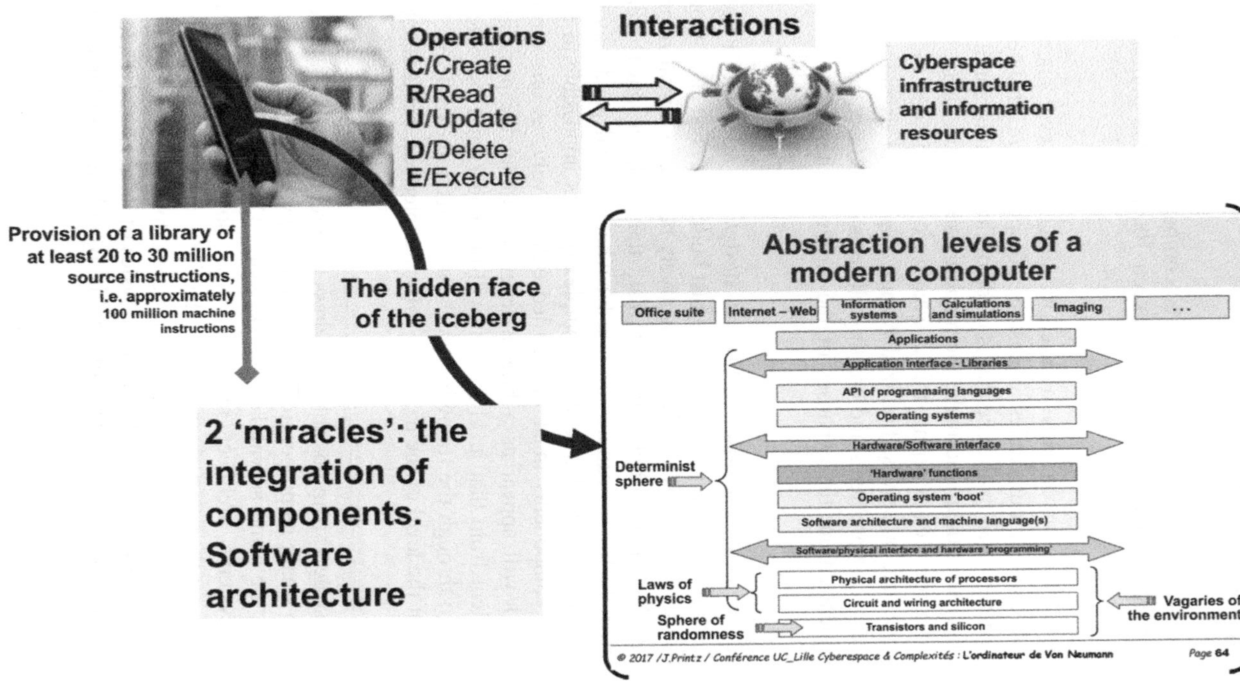

Figure 6.3. *From the PC to the smartphone (interior)*

The example of the smartphone allows us to see exactly what we should understand by emergence. In order to make an object like the smartphone "emerge", it must first be divided into elementary processes, continuing until the layer with the smallest of components is reached, that is, the nanometric transducers that will in the end cause the desired function to occur; the whole must then be gathered as a complete object, integrated, step by step and ready for use. The logic of emergence is a logic of integration, one that must be considered a constraint when dividing the object into the elementary processes. Conception and integration are therefore complementary activities, inseparable and indissociable, although entirely distinct, as is the case with the wave/particle duality in quantum physics. In the language of quantum mechanics, these two processes would be called "entangled" because, although distinct, the one cannot be defined nor exist without the other.

We can say that complex objects such as the smartphone are defined by their integration logic, also called "reference framework", which allows them to "emerge," that is, to simply live the life that their designers imagined for them! A good metaphor that can help us imagine the meaning of these "constructions" is that of the puzzle, already used by Alan Turing himself in one of his most famous texts, *Solvable and unsolvable problems*, written a few months before his tragic death[8].

6.1.1. *Chance and inaccessible numbers*

6.1.1.1. *Chance, a reflection of our ignorance?*

All the great mathematicians who have contributed to the development of the calculation of probabilities have had a take on chance, luck and/or good fortune, as was said in the time of Pascal and Fermat, the founding fathers of the discipline.

And only citing the French people, we are in the company of Laplace, Bertrand, Poincaré, Borel, etc[9].

We will provide a few reminders, taken from the work of Jacques Harthong, *Probabilités et statistiques*, an excellent book that can be downloaded, and whose passages on modeling are especially well presented; we will also provide a few examples from the book by Bernard Beauzamy, *Méthodes probabilistes pour l'étude*

8 The logic of these constructions also follows the definition that Feynman gave the term "construction" and the final remarks made by von Neumann. See the survey by Jack Copeland, *The Essential Turing*, Oxford University Press, already cited.

9 See, for example, Chapter 4, *Chance*, from his book *Science et méthode* (1909), regularly re-published, as well as the introduction from his lectures on *Calcul des probabilités* (1912), reprinted by Gabay.

des phénomènes réels, published by the *Société de Calcul Mathématique*, whose title is the program.

The calculation of probabilities was at first an intuitive theory, until the 1930s, when a definition of probability appeared that saw it as "circular", according to certain mathematicians. Axioms were introduced in 1933 by the great Russian mathematician, Andrei Kolmogorov, in his book *The Foundations of the Theory of Probabilities* (written in German with the title *Grundbegriffe der Wahrscheinlichkeitsrechnung*), in which he presents the axioms for calculating probabilities, a response to the sixth Hilbert problem, already mentioned above[10].

There are therefore two ways of teaching probabilities, either by continuously relying on intuition and the semantics of the real-world objects it represents, or as a formal game that is purely algebraic and completely disconnected from the real world. If we want to use it as a tool for modeling, it is the first method we must select. This was the pedagogical choice made by Harthong and Beauzamy.

In the preface to his lesson, Harthong remarks that:

> Science was once called natural philosophy and formed a whole[11]. Until the turn of the 20th century, every great mathematician understood physics as well as mathematics. Moreover: they were not separated. A veritable understanding of the concepts for calculating probabilities is impossible if we remain within the confines of one specialty or a prisoner to the "jargon" of one discipline.

As a definition of probability, he therefore chose that of Laplace, which he considered the best:

> The theory of chance consists in reducing all events of the same type to a certain number of equally possible cases, that is to say, so that we are equally undecided about their existence, and then determining the number of cases that are favorable to the event whose probability we seek. The ratio between this number and the one for all possible cases is the measure of this probability, which is thus a fraction whose numerator is the number of favorable cases and denominator the number of all possible cases (*Essai philosophique sur les probabilités*, 1819).

---

10 His body of work is considerable, for what interests us here. See: https://en.wikipedia.org/wiki/Andrey_Kolmogorov; and for more details, the two volumes of the festschrift *L'héritage de Kolmogorov*, published by Belin. Also see the work by J.-P. Delahaye, *Information, complexité et hasard*, published by Hermès.

11 Let us recall that the full title of the important work by Newton was *Mathematical Principles of Natural Philosophy*, published in 1687 and 1726 for the reference edition.

Hence the well-known formula $P(A)=\frac{\#A}{\#\Omega}$, where $\#A$ represents the number of trials that pertain to an event A, and $\#\Omega$ is the number of all possible trials. Identifying groups A and Ω must be completely rigorous (a "collectivizing relationship", in Bourbaki's terms) so that P has a meaning. To avoid having the calculations produce a meaningless result, we must be sure that the hypothesis of equiprobability, that is, the number of trials (we also speak of the *possibility space*) that constitute the event, is double-checked.

The language for calculating probabilities is very simple and includes the terms:

– trials, one of the possible results chosen by chance, without preference for any of them;

– cardinal, denoted by #, the number of elements in a finite set;

– possibility space, the set of all possible trials, denoted by Ω, and its cardinal $\#\Omega$;

– event, a set of trials, defined by a common property among its elements, denoted by A, and its cardinal #A;

– probability, see above, $P(A)=\frac{\#A}{\#\Omega}$.

The word chance is not part of the technical vocabulary, even though understanding it is necessary for modeling it correctly. It is part of the theoretical metalanguage, or perhaps its grammar. Without this understanding, calculating probabilities is a meaningless operation. For example, in many cases, we simply give a confidence coefficient of the kind "what is the probability that the Columbia shuttle will or will not have an accident?", according to what experts say, which corresponds to a number between 0 and 1, or a percentage; but that does not exactly make a probability and we know what tragic results this can have. Manipulating these coefficients with the rules of calculating probabilities is simply made-up mathematical nonsense. The term trial space suggests equiprobability, but we can never truly know what is equiprobable.

Beauzamy was even more categorical when he said:

> chance does not exist in practice: it is a practical word for masking our ignorance, for gathering within one word all sorts of causes, explanations, reasons, that we either do not know or do not wish to see. It is a synonym for laziness (*Méthodes probabilistes pour l'étude des phénomènes réels*, p. 21).

As a definition for the word chance, he explains:

> What we *cannot* explain: because we do not have the time, because we do not have the necessary technical or human means, etc.
>
> What we do not *know* how to explain: because we do not know the laws that react to the phenomena (they exist but we do not know them).
>
> What we do not *want* to explain: because we have decided that it isn't worthwhile, or perhaps because we refuse to seek the explanation (for example, for religious or philosophical reasons) (*Ibid.*, p. 36).

As for equiprobability, he was entirely categorical, providing as Rule 1:

> The probabilistic approach for incredibly rare and unrepeatable phenomena is completely devoid of meaning. These phenomena, by their very nature, do not justify a probabilistic approach. Any attempt to introduce one arises from intellectual dishonesty (*Ibid.*, p. 50).

The supposed independence of events, crucial for knowing which is the correct calculation rule to use in finding probabilities, is also a matter of good practice, or Rule 2:

> To ensure independence, it is not enough to ask if X and Y can each take on all the possible values (NB: we are dealing with random variables), whatever the value of the other may be. We must ask ourselves if X and Y have any "common factors", undergoing common influences (*Ibid.*, p. 97).

Since we all live on the same planet, we will have to know how to stop at a certain point and mark the territory (see the issue of *inside/outside* in systems engineering), something which requires good sense. And a little further on, we find Rule 3:

> In matters of independence, it is essential to question oneself. We could very well, at first, consider two random variables independent, just to revise this hypothesis later, if it turns out to be necessary. Thinking about independence is a form of art in engineering: it is crucial to know the protocols for measuring, how machines work, in order to respond to them. Independence depends on the precision of measurements.

IMPORTANT.– Here are some precautions to never forget, therefore, when exploring an unknown territory seen as complex where it seems like probabilities could be used.

Among these different rules, the most important one for the matter at hand in this book is the one that applies to rare and non-repeatable phenomena. Indeed, with natural or artificial complex systems, we are by definition in this category of phenomena.

Speaking of chance in terms of complex systems must be done very carefully. From the fusion of a human homo sapiens' ovum and spermatozoid, we will always obtain a homo sapiens man or woman, never an australopithecus or a chimpanzee. The combined genome will be unique, given the underlying combinatorics, but numerous variants can come from the stock provided by the chromosomes of both parents, some of which are unviable and would be eliminated during gestation through quality control mechanisms integrated into the embryogenesis process and the immune system, mostly unknown but that we can assume exist given the errors that accompany this highly complex biochemical process.

6.1.1.2. *How to model the "impossible"?*

In a Berkeley physics lecture in the 1970s, Figure 6.4 was used in the part of the course dealing with statistical physics to illustrate what is meant by a physical impossibility[12].

As for Harthong, he recalls a rule established by Borel:

> Events whose probability is infinitesimal, denoted by ε, never occur, which is to say that the probability $P = \varepsilon = 0$, strictly.

In our real world, such a scenario will never occur, even though we can calculate the probability, since this probability is an inaccessible number, infinitesimal, that corresponds to nothing in the real, physical world, even though we observe this reality at the scale of Planck units.

Figure 6.4. *Image of impossibility*

12 It appears in Volume V, Chapter 1, *Characteristic features of macroscopic systems*. This lecture can be downloaded from the Internet.

As Boltzmann would say, it defies imagination, but we have to get used to it!

The real world is not a random construction, materialized from a void. There are regularities everywhere, beginning with the atoms themselves, veritable constructions made of "quarks". The rules of emergence must be respected, for they simultaneously sculpt the rules of the physicochemical world, the H1 hierarchy where information plays a minor role, H2 hierarchy rules, the hierarchy of the living, and H3, the hierarchy of computerized complex artificial systems, which, in fact, make up just one, the informational hierarchy (Figure 6.5).

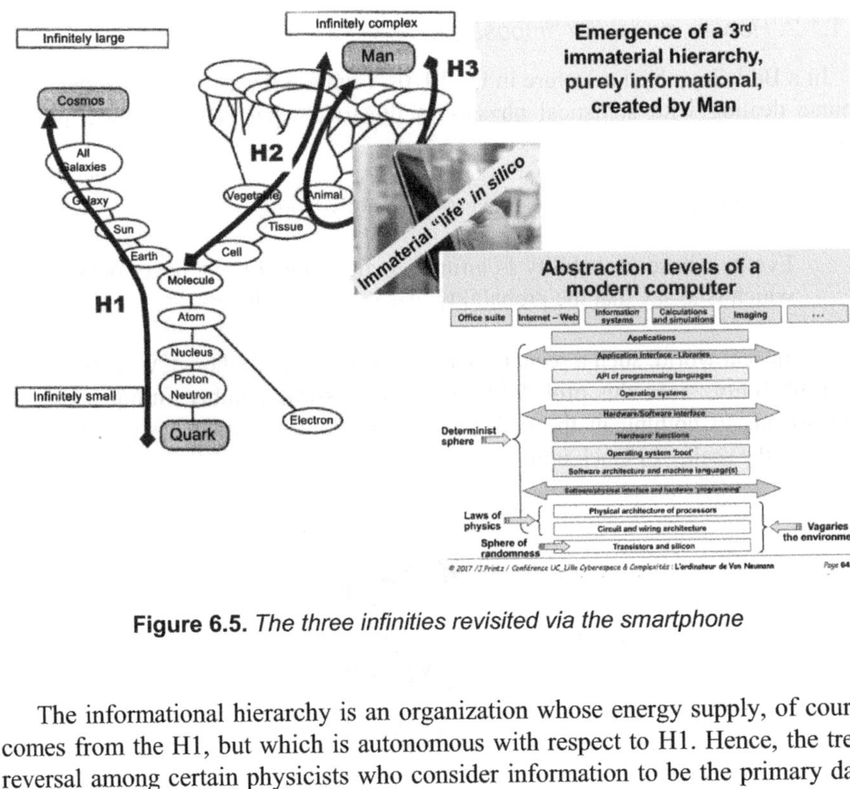

Figure 6.5. *The three infinities revisited via the smartphone*

The informational hierarchy is an organization whose energy supply, of course, comes from the H1, but which is autonomous with respect to H1. Hence, the trend reversal among certain physicists who consider information to be the primary data, for example, John Archibald Wheeler and his "it from bit" formula, which ought to be re-examined in further detail[13].

13 This is the title of a chapter in his book, *Geons, Black Holes & Quantum Foam*. In it, we also find this surprising assertion: "'More is different' may have something to do with 'it from bit.' The rich complexity of the universe as a whole does not in any way preclude an extremely simple element such as a bit of information from being what the universe is made of. When enough simple elements are stirred together, there is no limit to what can result."

Regarding the role of information in the living, let us refer to the work by Jean-Louis Dessalles, Cédric Gaucherel and Pierre-Henri Gouyon, *Le fil de la vie – La face immatérielle du vivant*[14]. This very interesting book opens perspectives on relevant similarities between the H2 and H3 hierarchies and the corresponding fields of engineering.

6.2. The organized objects of the 3rd infinity

The organized objects of the 3rd infinity are of two kinds: those that come from the sphere of life, like Laughlin's chicken and those that come from finalized human activity, such as the smartphone (Figure 6.6).

We can assuredly attempt to make a parallel since the two rely on the same logic: that of information and the languages associated with it[15]. We are well acquainted with the blueprint for building an object such as the smartphone, as well as its *modus operandi*, that is, its programming, and the reference framework that defines it as a system/object in its context of use[16]. We do not know the rules for the chicken's ontogenesis, nor indeed that of any other living organism, even though biologists naturally have some ideas and hypotheses on the matter, even though that means knowing only the number of genomes and the indispensable sub-systems for its survival such as the immune system[17].

The fundamental question is knowing if the one sheds light on the other, and in which order. Unless we are dealing with a systemic cycle, like the one for which Wiener was the first to point out the importance in his book *Cybernetics or Control and Communication in the Animal and the Machine*. At our meso-/macroscopic scale, we are under the impression that the world is stable, or at least that certain things are[18]. But the reality is that everything is unstable, everything vibrates and everything is cyclical. It is only at absolute zero that the world becomes immobile, what physicists call the fundamental state, obtained in the Bose/Einstein condensates. The appearance of stability comes from our tools for perception, from our senses, which means that beyond a certain threshold, the instability of reality disappears.

14 Odile Jacob, 2016.

15 Firstly, see the book by Jean-Paul Delahaye, *Information, complexité, hasard* (Hermès, 1994). His small book from 2009, *Complexité aléatoire et complexité organisée*, is a good summary.

16 Let us recall here the work of Simondon, already cited.

17 See the conclusion.

18 Until the time of Galileo/Newton, the sky remained the "sphere of fixed objects", inherited from Aristotle and the Greeks.

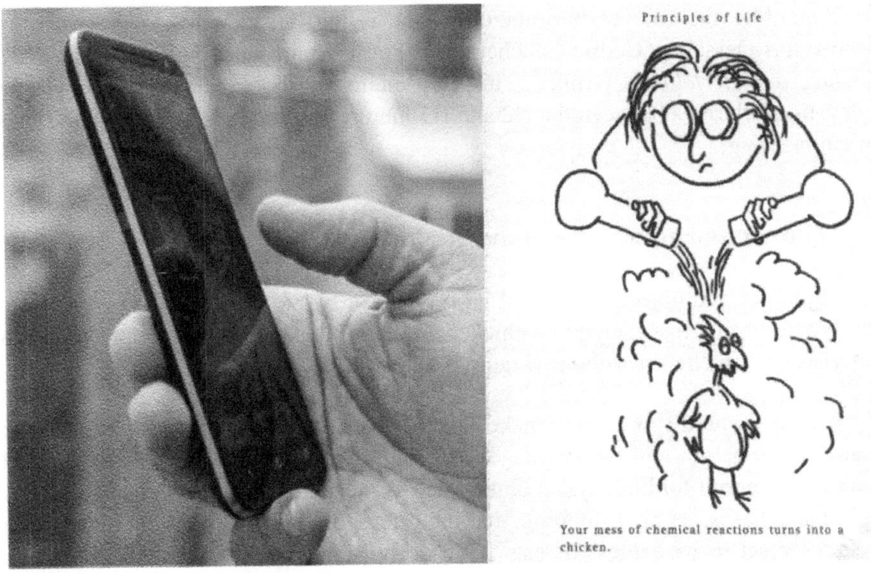

Figure 6.6. *Varieties of complex systems*

6.2.1. *Engineering hazard*

However, strange and paradoxical it may seem, there is a science and an engineering of chance[19]. In nature, many entirely regular phenomena occur, without which the world would be unstable and meaningless, to use the expression by Thom (already cited). There are also irregular, unique phenomena that are never repeated in exactly the same way, therefore irrelevant to the calculation of probabilities, but which reinforce the regularity of certain other phenomena: therein lies the paradox.

We can elucidate this matter by giving some examples from the engineering of complex artificial systems, leaving the door open to the engineering and ontogenesis of biological systems where, it is clear, chance also plays a role in survival. Here, we are referring to the joint work done with Professor Philippe Kourilsky on the immune system, which will serve as a conclusion[20].

---

19 It is the science of approximations with its so-called Monte Carlo method that is the best example, created entirely by von Neumann and his friend Stanislas Ulam, using experimental mathematics whose applications are important to multiple fields.

20 See the conclusion to this book.

As for the practical use of random phenomena, the *Poilus* or infantrymen of World War I understood that the best protection when under fire was to take cover in the craters made by the first shells; they had noted that when fired, projectiles never fall in the same place twice.

In the sciences and technologies of information and communication (STIC), chance has been baptized as "noise", and it is congenital with their birth, or we could almost say intertwined. "Noise" is everywhere, like the meaninglessness that Thom spoke of; noise means nothing; if not, it is no longer noise! At the foundations of information science, we find the fundamental works of John von Neumann, *Probabilistic Logics and the Synthesis of Reliable Organisms from Unreliable Components*, Claude Shannon, *Mathematical Theory of Communications*, and Norbert Wiener, *Cybernetics or Control and Communication in the Animal and the Machine*. Without these theoretical contributions, profound and subtle regarding the role of randomness, we would not have had computers nor information science.

We will examine two exemplary situations, omnipresent in STIC, in the form of two questions: (1) How do we create randomness? (2) How do we use randomness, or "noise", as defined by von Neumann and Shannon, to create stability?

6.2.1.1. *First situation*

One of the most frequent operations in STIC is searching for previously archived information in a memory. This is what search engines do constantly, but the issue has always existed, including with mechanographic materials, well before computers. The search mechanism is entirely general and works on the basis of recognition, which starts with perception. We see something, through the intermediary of our senses, and we must find, rather quickly, the information associated with this perception. In the jargon that some people like to use, this is what we call pattern matching[21]. For example, using the name/address of a taxpayer, we can find their fiscal information; better yet, using the image of a face captured by a security camera, we can find out if it is on a database of wanted individuals. Such examples are innumerable.

If we take the case of names/addresses, the initial data is anything but random. The names are not just any names, and often reflect a characteristic of the person. Some names are more frequent than others. If we only use a name to perform a search, the disparity in names will create a disparity in the search time that could turn out to be fatal in certain situations. If we believe the statistics of patronyms, the most common name in France is Martin, with 4,158,286 occurrences, while there

21 This is also what networks of "neurons" used in deep learning do.

are only 3,093 people with the name Printz[22]! In order to create a balance in the time and resources needed for a search, a system that erases imbalances must be invented, and this is where randomness will first be useful. We have seen that the problem of arranging items in boxes translates into combinatorials that create inaccessible numbers. To avoid ending up with boxes or aisles that are completely saturated while others are empty, we must find a naming mechanism, that is, a name attribution mechanism that erases the initial disparity; rather than being called Dupont or Schmidt, we can assign codes to names, for example, 1608FAF9FA and 8376F86817 – here, in a hexadecimal alphabet {0, 1, 2, …, 9, A, B, …, E, F}, but we could just as well have chosen the four letters of the genome, ATCG, since by assigning them 2 by 2, we create 16 combinations.

This type of correspondence is created by hashing algorithms, a term related to common expressions such as *hashtag* or *hashcode*. These algorithms were invented as early as the 1950s–1960s and have since been in common usage in the computing community. All systems offer the service of randomly generated names. The computer I worked on the most had a command called HASH which would take a chain of up to 256 characters and return a hashcode in 32 bits, which in turn would lead to a set whose cardinality was 256^{256} or $3.24 \ldots \times 10^{616}$ (an inaccessible number) over a set of 32 binary digits, thus 2^{32} or $4.29 \ldots \times 10^{9}$, which is a normal number, in other words, exactly 4,294,967,296, yet still too large to represent the French population (estimated to be 67 million) or the file listing taxpayers (slightly less than 20 million). The corresponding Galton triangle would only have 32 levels, with most of the columns empty, for a population like France's. To stay within the confines of the blueprint in terms of performance, the hashcode still needs to be reduced by dividing it by a random prime number, so as not to perturb the randomization – let us say something around 100,000 boxes, taking, for example, the number 4,297, which makes for a total of 999,527 boxes, each of which contains the information for six or seven French people. By doing this, the time required to access the information will be the same for everyone. Hence, the sequence of stabilized operations: (1) observation, (2) hashcode calculation, (3) sequential or associative search of the corresponding storage box and (4) access/entry of relevant information.

It is thus clear that randomness, used in this manner, works to simplify the recognition mechanism by creating regularity: if the algorithm is good, all searches will have the same cost. This is a fundamental point since letting chance follow its "natural" course would have led to: (a) an immense disparity in the costs of a search, leading to an intrinsic, uncontrollable instability and obvious survival issues and (b) a gigantic waste of resources where the majority of storage boxes remain empty,

---

22 See: http://www.geneanet.org/genealogie/1/Nom-de-famille.php.

making the system radically uncompetitive[23]. We must never forget that we live in a world where energy limits are everywhere, and that we are ourselves individually "limited". There is a profound analogy between what has just been presented and Shannon's first theorem on the existence of omnipresent economic codes in information science.

We do not know if comparable mechanisms exist in the engineering of life, but it is absolutely clear that powerful addressing and recognition mechanisms are needed to organize and find the information that helps organisms survive, including the most rudimentary of them, with a truly stupefying economy of means. To use the example given to me by my colleague Kourilsky, coagulation starts precisely at the site of a wound, and nowhere else, for inappropriate coagulation in the heart or the brain would assuredly lead to death. Bringing chance into situations of this type is ipso facto the kind of anti-scientific laziness identified by Beauzamy and vigorously denounced by Thom and Chumbley (1983) in their article "Stop Chance! Silence Noise!"[24], according to which, for all explanations, we content ourselves with denying a problem in order to hide our ignorance, somewhat like the physicists at the end of the 19th century, who refused to take the atomic hypothesis seriously under the pretext of not being able to see atoms.

6.2.1.2. *Second situation*

In widely used systems such as the smartphone, which integrate billions of nanometric components and millions of lines of code, and even more so in laptop computers and tablets, these great *data centers*, the kind of noise of which Shannon spoke, are everywhere, whether in the form of natural radioactivity, thermal agitation of the materials used or the influence of magnetic fields, including magnetic storms like the ones that caused an enormous failure in power distribution in Canada on March 13, 1989.

Concretely, this means that in the memory component of a computer, let us say of 32 bits, one of the bits will go from 0 to 1, or vice versa, in a random manner. Let us recall that what is seen as a bit by a programmer is, in fact, a small electronic circuit (an oscillator) made of a few components that are fed by an electric current. Everyone knows that if we cut the power, the content of the memory is lost. It is therefore easy to conceive of a random energy supply coming from the environment, in any form; for example, a simple shock is capable of changing the energetic state of the area where the bit in question is located. In this case, the encoded information using this bit no longer makes any sense. These random distributions, whose external manifestation is called an error, were (and still are) the number one problem

23 In biology, a non-competitive organism is eliminated by evolution and the competition between species.

24 Thom, R. and Chumbley R.E. (1983). Stop Chance! Silence Noise! *SubStance*, 12(3), 11-21.

with the first computers. Today's users will surely have forgotten that until the start of the 1980s, before the time of personal computers, computers took up a lot of space, and to make them work, a lot of energy was needed, along with powerful means of refrigeration, but this was not enough to avoid the effects of "noise".

If, for any reason, following a hazard of the environment, a memory word of, let us say, 16 bits in older technologies, has one of its bits change state, the first operation needed is to detect the change of state while the memory is being read. For that, in keeping with the fundamental work done by von Neumann and Shannon, we add another redundancy bit to the original 16 bits, which is a 0 if there is an even number of 0s among the 16, and a 1 in the case of an odd number; this is what we call a parity bit. This way, if one of the bits from the 16 changes state, all that has to be done is to recount them at each use and compare the result to the parity bit. If the count is correct, we can continue.

In doing this, there are two problems. If, instead of 1 bit, there are 2 bits that change state, then the error is undetectable by a parity control; this could be 2 bits from the 16 or 1 bit from the 16 *and* the parity bit. Nevertheless, the corresponding event is a double event that simultaneously affects one memory zone. If we understand the nature of noise, we can calculate the probability of such events. In the computers of the 1970s, up to the start of the 1980s, the only parry for this kind of situation consisted of taking periodic photographs of the memory (which was much smaller than current memories), let us say every hour, and if an incoherence was detected – since a visible incoherence will always end up appearing – the task was restarted using the most recent coherent image. If the phenomenon is truly random, it will not be repeated, and the work can restart and finish as planned.

If, however, the disruptive phenomenon is not random and demonstrates a reproducible and repetitive character, then we have a good chance of detecting it using the classic trial and error approach with more and more detailed targeted observations, analyzing it so as to identify the cause and control it, but not in real time. This was a problem until the 1990s, a period when "real-time" specialized computers began to disappear.

In the 1980s, STICs underwent a technological revolution thanks to the generalization of transistors and computers that were entirely *made of transistors*, which led to the success and glory of Digital Equipment and its charismatic president Ken Olsen, a famous designer who has now passed away, and especially thanks to the appearance of a new type of transistor using what was called CMOS technology, which is characteristically very simple (all other things being equal, since making transistors is anything but simple), has low energy consumption and does not need much silicon to get the desired effect[25]. Since they are small, very small even, it became possible to place lots on the same substrate. This was the so-called LSI wave, and then VSLI (*Very Large-Scale Integration*). It was the period when, for the first time, it was possible to put the equivalent of a central unit on just one circuit, for example, the Motorola 68000 circuit, which contained 68,000 transistors; this was an LSI. VLSI appeared in the 1990s, when the threshold of about a dozen million was surpassed. Today, in the latest technology, we have what is called SOC (System On Chip), which contain billions of transistors, for example, the 22-core Xeon from Intel with its 7.2 billion transistors, or even the one in the NVIDIA QUADRO GV100 computer and its 21 billion transistors (25 million per mm^2).

In one decade of continuous progress, the transistor has gone from the status of a rare resource to that of an abundant resource. This change has had a twofold effect on reliability. The word memory has gone from 16 to 32 bits, then 64 and even 128 for memory buses, which enable communication between the memory and the unit of functional calculations called the central unit or the CPU.

First consequence: it has become possible to implement error correction codes, or ECCs in the jargon, a theory developed by Shannon, in a systematic way. This was Shannon's second theorem, that of the channel with noise; a truly fundamental theorem since it states the property that whether or not the nature of disruptive noise is known, there are codes that not only can detect the errors, but also correct them. An immediate practical consequence is that parity errors disappear from the application programmer's field of vision, but they naturally still exist and are simply masked. Each piece of information has an associated code that detects and corrects. For our 32 bits, we have ECCs of 7 or 11 bits, and even more for external hard drives or telecommunications. Thanks to these ECCs, NASA was able to remain in contact with its *Voyager* probes from the end of the solar system.

It is important to understand when implementing these mechanisms that everything depends on the random nature of disruptions. We saw that with inaccessible numbers, cardinality is such that even a duplicate is an inaccessible number.

Repetitiveness, the fundamentally non-repeatable character, is therefore a phenomenon that does not belong in the world of randomness, which is what is useful about it. But we must still make sure of this, by constantly monitoring everything that happens inside and outside of a system (again, the aforementioned issue of *inside/outside*).

25 See: https://en.wikipedia.org/wiki/CMOS as well as Chapters 2 and 5.

Second consequence: an abundance of transistors makes it possible to integrate a kind of dual computer in the computer, whose role is only to monitor the properties that are characteristic of the nominal state of the functional computer, thanks to a well-dosed redundancy – this is the major challenge for architecture and the person designing it – as the addition of redundancies must not create other problems. In computers from the end of the 1980s/beginning of the 1990s, the reliability of hardware made a giant leap, something which could lead us to think that errors disappeared as if by magic, when really they were only being compensated for and computers were being constantly monitored. The monitoring function was named the "demon" by some, in reference to Maxell's demon[26]. Others, who were more neutral, called it the Service Processor. It was the concrete execution of a famous expression used by von Neumann: "who are the guardians"[27]? In a current SOC, approximately 15–20% of the available transistors are used for monitoring and/or correction, without counting the "online" test data, which must of course be available in real time on the SOC.

As for the software, the nature of the issue is the same: detect and correct anomalies. But the corresponding engineering is purely functional and abstract, radically different in its structure, for it no longer deals with materials or physical phenomena. Errors are made by programmers in programs whose size is counted in the millions of source instructions written by them.

To illustrate this, we can imagine what happens in a *Data Center* such as the ones that the GAFA companies set up more or less everywhere, but especially in colder regions where the heat produced can be recovered (and create a larger profit margin by selling it) using hundreds of thousands of elementary CPUs for the job. These are probably the most complex "computers" ever created by humankind. Monitoring is restricted to a system that is supervised by human operators in the system platform itself[28].

Keeping within our theme of inaccessible numbers, and touching upon the way we must approach this in order to avoid getting lost in it, we must imagine that each of the software modules operating on the thousands of CPUs in the data center each contain a work zone for managing their own data, a zone that will be connected to the zones of shared memory so they can communicate with one another. The

26 Such as the designers of the MULTICS system, developed at MIT and with whom I was lucky enough to collaborate.

27 In his thesis "The general and logical theory of automata", *Complete works*, vol. V.

28 See: https://en.wikipedia.org/wiki/Mercury_Interactive and https://en.wikipedia.org/wiki/Tivoli_Software. In computing jargon, it is called *Systems management*, see: https://en.wikipedia.org/wiki/Systems_management.

synchronization of activity and sharing are executed with synchronization mechanisms like those represented at the bottom of Figure 6.7.

This comes down to facilitating the passage of a traffic flow through a network, hence the traffic lights, semaphores, messages and mailboxes, as if managing all traffic and communication in a city or port.

If everyone can communicate with everyone, two by two, three by three, etc., we are left with a set of parts, or 2^N, with N being the number of modules. If each one of them leaves traces of their execution in the shared memories, then these states will recombine with the previous states, now creating a combinatorial that grows like $2^{2^{\cdot^{\cdot^{2^N}}}}$, depending on the number of steps made in the calculations. If the system is not monitored in real time, there will be no chance of finding what might have caused any anomaly or other situation that users wish to examine.

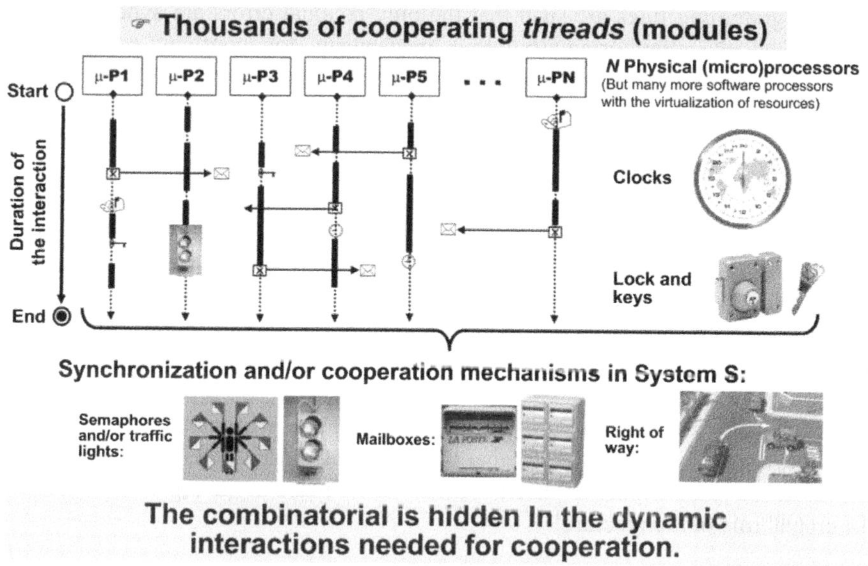

Figure 6.7. *The world of programmers*

In a *Data Center*, these are tens of thousands of software modules called, depending on the function, processes, threads or transactions, all being executed in parallel and interacting, immediately creating an inaccessible number of situations. If the modules share certain resources, the resource access requests must be managed, since if, for example, M_A requests resources R1, then R2, etc., in this order, and M_B, in parallel, requests R2, then R1, we find ourselves in a situation known as a death grip since M_A is in possession of R1 and is requesting R2, but R2 is already preempted by M_B which is requesting R1. The system becomes stuck. The only possible recourse is the surveillance system S_{Surv} that must note the conflict and arbitrate to unblock the system[29]. In theory, an execution flow containing N instructions can be interrupted N times, which is why for this flow alone there are $N!$ possible configurations, and for the set of n flows interacting together in Figure 6.7, there is an inaccessible number whose order of magnitude is $N_1! \times N_2! \times \cdots \times N_n!$.

Incidentally, we should note that if programmers do not scrupulously respect the rules for sharing, there is no chance of this working; we must therefore trust them! Ethics sneaks into the world of programmers, which explains why ethical codes like that of the IEEE or the ACM exist[30]. There is no escape, especially not for AI or Big Data, whose underlying complexity can make them inoperable because of combinatorics and inaccessible numbers generated in this way.

If programming rules are such that the only residual errors are synchronization errors, then the monitoring/arbitration mechanism for conflicts can organize rematches, for example by temporarily cancelling the job already done by the module/transaction that caused the conflict, which frees the first transaction, and replanning the faulty transaction for a bit later, resupplying it with its job data. From an ethical point of view, the programmer must ensure that their program is repaired if they have all the resources they need at their disposal. In the case of a conflict, they defer to the transaction monitor, which arbitrates conflicts with the specific condition of respecting common rules. These rules work as a "self" that defines one criterion for belonging to the community – this is a constant in computing – since everyone accepts its rules. The acceptable conduct, following a probabilistic PAC logic, is based on this engineering of chance that we just briefly evoked.

NOTE 6.2.– My colleague Kourilsky pointed out that the immune system contains lymphocytes called *natural killers* that will, without warning, "kill" any cell that is not identified as belonging to "itself" or that has no place in the organism – showing the crucial role of architecture!

In such systems, it is clear that chance has no place simply because it is a deadly risk. If the designer lets chance in, they are already "dead" because they will immediately be drowning in the insignificance of inaccessible numbers. Everything is built, integrated and validated, step by step, stage by stage, capitalized on and reused, waterproofed with security airlocks that make up the interfaces, where they can "observe" what goes on via the monitoring system (in this case, a completely

29 These mechanisms were analyzed in our previous books, *Architecture logicielle* and *Architecture système et complexité*, Vol. I.

30 See: https://www.acm.org/about-acm/acm-code-of-ethics-and-professional-conduct.

abstract computer), doing this until they reach the top of the stack, which will allow everybody using a social network to contact their followers, forgetting what happens beneath, but with total trust in the people who designed and developed it, such that everything relies on the intelligent integration of the "more is different" logic and the PAC logic[31].

NOTE 6.3.– In high availability systems, the architecture can be completed with redundant systems using majority votes with so-called *N*-version programming, used in highly critical systems to introduce controlled randomness, known for ensuring operability[32].

6.3. Facing the immensity of the infinitely complex

We have all had the experience of walking through the infinitely complex when we learn our first language since, understandably, the first contact a newborn has with the outside world occurs through the intermediary of their mother, including in gestation. This fundamental lesson is first learned through imitation, and then progressively, from about the ages of 5–7 years, the child acquires the capacity to describe situations that are specific to their experiences, and even ones they "invent", and to communicate these to their immediate social environment: parents and grandparents, uncles and aunts, brothers and sisters if they are not the first-born, and cousins, etc. From a passive receiver and imitator, the child becomes an emitter and active creator through a process like integration[33]. This is also the age when they begin to master reading and writing, in whatever form, alphabetic or ideographic.

This linguistic capacity, unique in the animal kingdom, carries a name: it is a grammar. Even though we are not fully aware of it, mastering a grammar constitutes a first level of autonomy that everything else relies upon, which gives a child and then an adolescent the ability to interact efficiently with their spatiotemporal environment by memorizing it. Language clearly plays a primordial role in this system of symbolic representation of the outside world. Learning a language is a game whose rules are the grammar; its complexity by far surpasses the rules of chess or the game Go, presented by some as a measure of complexity – a rather naive proposition, to say the least.

31 In reference to the book by Leslie Valiant, cited above.

32 See the DO-178B norm, https://en.wikipedia.org/wiki/DO-178B, *Software Considerations in Airborne Systems and Equipment Certification.*

33 See Chomsky's writing on grammars; also see Piaget, *L'équilibration des structures cognitives* (PUF), and his work on developing intelligence.

We are very familiar with the dynamic of this learning: slow at the start, followed by an accelerated phase of exceptional speed, then a plateau of maturity. It is the curve of the logistic function, called an *S*-curve, because of its characteristic sigmoid form. If C is the competency acquired at time t, starting with an initial innate capacity C_0, we can model it as a differential or an analytic solution, as follows:

$$\frac{\Delta C}{\Delta t} = (k - \varepsilon \times C) \times C$$

whose integration gives:

$$C(t) = \frac{kC_0 \times e^{kt}}{k + \varepsilon C_0 \times (e^{kt} - 1)}$$

We can simplify it by taking the first linear term of the series expansion of e^{kt}, that is, $1 + kt$, which gives the simplified equation $C_L(t) = C_0 \frac{1 + k \times t}{1 + \varepsilon C_0 \times t}$. In both cases, when t grows, the asymptote is the same: $\frac{k}{\varepsilon}$; this does not depend on initial competence, but only on the factors of acceleration k and decay ε[34].

In the later stages of their development, the adolescent who has become a young adult will discover two other families of language along with their grammars, from which they cannot escape[35].

The first is the language of natural phenomena, whose grammar, as Galileo told us five centuries ago, is the language of mathematics – a language that lets us interact with nature while revealing its most inner workings, making interaction without destruction possible. This is the language of science and the techno-science of engineers, a language that allows us to understand complex objects such as the smartphone, why and how "it works". It is a universal language, independent of the European cultural context that gave birth to it. All cultures have established a corpus

34 For a more researched study, see the book by Vito Volterra, *Théorie mathématique de la lutte pour la vie*, reprinted by Gabay; or for something more recent, that of Kowak, *Evolutionary dynamics* (Harvard University Press).

35 For developmental theories, see Jean Piaget, in particular *L'équilibration des structures cognitives*, which gives a general qualitative theory of development, and his numerous other books, including the three volumes of the *Introduction à l'épistémologie génétique*. On grammar, see Noam Chomsky.

of scientific knowledge, often grandiose as is the case in China[36]. But only Europe, thanks to the invention of the hypotheticodeductive method, or the scientific method, used by the Ancient Greeks and vigorously updated between the 12th and the 17th centuries, has given it this very specific characteristic of universality that in fact surpasses it[37].

The second language is that of our metaphysical queries, induced through our models, formulated by the Greeks with the precept "know thyself," engraved on the architrave of the temple of Delphi, and the questions that accompany this knowledge, the fundamental questions that characterize the path of human existence: who am I? Where do I come from and where am I going? Why does evil exist? What will there be after this life?

These questions have a common source: the quest for meaning, which has always been significant in the heart of humans, whatever their culture, for the orientation they give their lives depends on the response to these questions. This is the language in which faith and reason, physics and metaphysics are articulated, structuring the entire history of thought in what we call the Occident, in contrast to the Orient, which has a different reference framework that has not given science the same importance. The metaphysical foundation interacts with physics, or simply nature, as Aristotle and the Stoic philosophers taught us, but the one can disguise the other or otherwise galvanize it.

Closer to our time, *The Key to Dreams*, the last book – available on the Internet – by the mathematician Grothendieck, one of the greatest mathematical minds of the 20th century, and perhaps even the entire history of mathematics, fits perfectly into this approach[38]. This was also the case for Weyl and von Neumann during a lecture given at Columbia University in 1954, *The Unity of Science*.

For a young adult of the 21st century, the complexity of the world as they discover it is truly frightening, capable of making their head spin if they are not

36 See the 37 volumes of the encyclopedia, *Science and civilization in China*, directed by Joseph Needham; in particular, Volumes II, III and VII-1.

37 See the recent book by Michel Blay, *Critique de l'histoire des sciences* (CNRS Edition, 2017) on the history of this unique transition; as a complement, see the ones by Arthur Koestler, *The Sleepwalkers: A History of Man's Changing Vision of the Universe* (Penguin, 1989), and by Pierre-Noel Mayaud, *Le conflit entre l'astronomie nouvelle et l'écriture Sainte au XVIe et XVIIe siècles – Un moment de l'histoire des idées autour de l'affaire Galilée* (Honoré Champion, 6 volumes, 2005).

38 See the conference given by Laurent Lafforgue at the 2017 ENS seminar, *Lectures Grothendieckiennes*; http://savoirs.ens.fr/recherche.php?rechercheTerme=Lectures+grothen dieckiennes.

well-prepared for the shock of reality. They will have to learn to move, act and build their life in a world they will never entirely master. They will have to learn to trust and to create trust. We can certainly find the combinatorial of what we see, but the problem is the meaning of this combinatorial. In a world where "everything is connected," how can we act without disrupting the edifice, how can we distinguish between what is beneficial to the development of the individual and the collective they belong to, and what is not, or what puts us on the wrong path? How many degrees of liberty do we really have? How do we put the pieces of this gigantic puzzle together, and how do we continue it? And to do *what*?

The science of complexity, in particular information science, which arose in the 1950s, shows us that the sphere of nonsense and irrationality is an infinitely more populous space than that of good sense and reason. There are infinitely more real numbers than rational numbers. To illustrate this quest for sense, we will use the "meaning map" created by the mathematician René Thom in his book of interviews, *To Predict is Not to Explain*, which can also be easily found on the Internet (Figure 6.8)[39].

We could write pages of commentary about this rather striking image, but what Thom tells us, more or less, is that our main problem is not that of truth or facticity, which can be seen at the bottom of the drawing, but that of the "sea of insignificance". The combinatorial that is not regulated by some finality leads us directly into this sea of insignificance, just like Forster's magnets[40]. Having an objective is what creates meaning, or to use Wittgenstein's aphorism, the "what is it for".

Faust, in the play by Goethe, wishes to leave behind the obscurity of hollow words, which leads him to the fatal pact with Mephistopheles[41]. If we let go, even just a little, without acknowledging the combinations we are always tempted to make, without integrating knowledge and validating the pertinence of this integration, the combinatorial will lead us directly to insignificance and nonsense. Ludwig Wittgenstein, in his *Tractacus*, written in 1921, and well before him, William of Occam, had already warned us in Aphorism no. 3328: "If a sign *has no use*, it has no meaning." Thom, with his theories of morphogenesis and semiophysics, imagined himself, or dreamed of being Aristotle's continuator. It is interesting to reread his lecture from the colloquium organized by the Pontifical Academy from October 27–31, 1992, on the theme of "The Emergence of

39 Already cited.

40 See his principle of "order from noise", vigorously disputed by Thom in *Halte au hasard, silence au bruit* (Les Débats, 1980). For more details, but no dispute, see the books by Dupuy, *Ordres et désordres* (1982) and *Aux origines des sciences cognitives* (1994).

41 Part One, *Night.*

Complexity in Mathematics, Physics, Chemistry and Biology", entitled *L'émergence des structures*, which is from the period of his interviews at France Culture with Émile Noël, but in a more, let us say, censured version[42]. As we indicated at the start of Chapter 4 via a remark by Feynman, we must understand complexity using a plurality of models like those presented in Chapter 1.

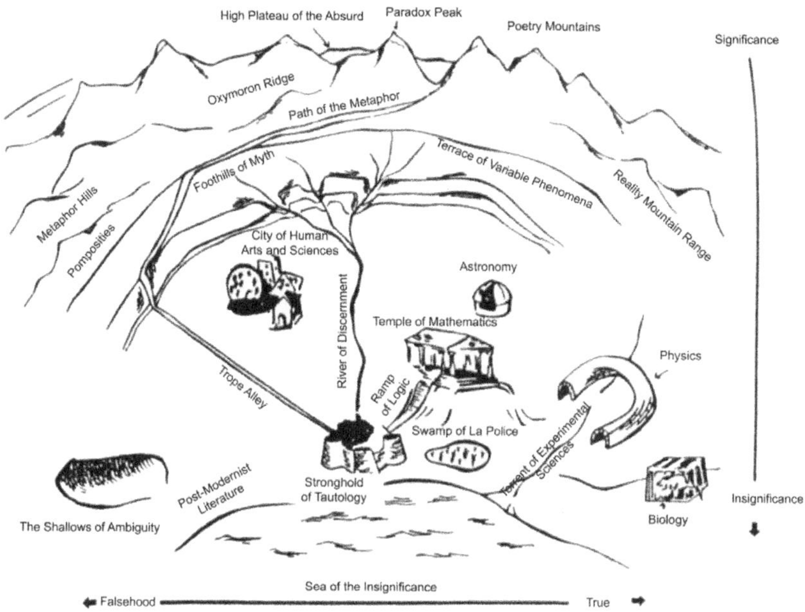

Figure 6.8. *René Thom's meaning map*

We find the same kind of warnings in the work of the mathematician and epistemologist Ferdinand Gonseth and in the work of the philosopher and logician Jean Ladrière, and in a general way, in the field of analytic philosophy, popular in the English-speaking world, which Jacques Bouveresse, a professor at the Collège de France, tried to bring to France, unfortunately without very much success.

IMPORTANT.– The arguments made in this second part of the book show, if need be, that to make progress through complexity without immediately slipping into a gulf of nonsense and insignificance, we must never resort to chance. We must proceed by building and integrating according to a probabilistic PAC – the very object that ensures that systems function – based on the correct properties of randomness discussed here.

42 This can be easily downloaded from the Vatican website.

The complexity of biological life is certainly not by chance, even though the mechanisms that underlie it are still mostly unknown to us. It would be prudent, in the sense of the cardinal virtue and/or Aristotle's *phronesis*[43], to avoid any reductive ideology such as the one used by physicists at the end of the 19th century when they refused to study atoms under the pretext that they could not be seen. We know how that story ended.

As for the complexity of artificial life, that of our artifacts and our systems like the smartphone, it owes nothing to chance and everything to the constructive intelligence of the people who were and continue to be its creators. We control all of the stages and how "it works" even though we are still very far from being able to compare it to the feats of bioengineering. The underlying knowledge in complex systems engineering that makes the emergence of these almost miraculous objects possible, for someone who does not know how they are made, is of a fundamentally linguistic and grammatical nature; language and metalanguage we could say, the basis for a reference framework without which collective action is impossible. This is what underlies the additive logic of the principle of *more is different*, this manifesto of emergence, that is, of the integration proposed by Anderson, this logic without which the functional stack of the smartphone would not only be unthinkable but also entirely unexplainable.

To conclude this study, let us finish with a citation from the mathematician Alexandre Grothendieck, a real genius in mathematics, a reformist of the field of algebraic geometry where, like Hilbert, he followed in the footsteps of Euclid, and who passed away in 2014, in anonymity. In his autobiography, called *Récoltes et semailles*, never published but available on the Internet, a small text can be found that clearly illustrates the method that anyone who wants to approach the systems that are all around us seriously, those which we create and those which are given to us by nature, ought to implement (p. 48)[44]. This short text echoes some of what can be found in the work of Turing and von Neumann:

> The structure of something is not something we can "invent". We can only expose it to the light of day, patiently, humbly getting to know it, "discovering" it. If there is creativity in this work, and if we happen to do the work of a blacksmith or a relentless builder, it is not at all to "fashion" or "build" any "structures." These have not waited for us to exist, or to be exactly what they are! But it is to express, as accurately as possible, these things that we are discovering and surveying, this

43 See: https://en.wikipedia.org/wiki/Phronesis; translated as "intelligence, wisdom".

44 Only available on the Internet. It is a large volume of 1,000 pages, dense but excellent, in which Grothendieck dissects the mechanisms for creating mathematical objects that allow us to "generalize in order to simplify".

> structure that is reluctant to reveal itself, that we are trying to examine, understand, perhaps through a still-stammering kind of language. We are thus constantly pushed to "invent" a language capable of expressing, in more and more detail, the intimate structure of the mathematical object, and to "build" the "theories" that are meant to explain what has been apprehended and seen, step by step, from start to finish. There is, in this, a constant back and forth movement, uninterrupted, between the apprehension of things, and the expression of what has been apprehended, using a progressively refined language that gets recreated as the work advances, constantly under the pressure of immediate needs.

In light of this enlightening text, I would simply like to present another text, this time from the mathematician and logician Alan Turing, concerning the insufficiency of reason – he who was one of the great masters of reason and who proved as much by decrypting the Enigma machine's code. Certain historians say that in doing this, he shortened World War II by two years, saving millions of lives. Here is what Turing himself had to say in one of his last pieces of writing:

> The results which have been described in this article are mainly of a negative character, setting certain bounds to what we can hope to achieve purely by reasoning. These, and some other results of mathematical logic, may be regarded as going some way towards a demonstration, within mathematics itself, of the inadequacy of "reason" unsupported by common sense[45].

This text, which unless I am wrong is not available in the public domain, was transcribed and presented in *The Essential Turing* by Jack Copeland (2004), with a very interesting commentary from the author.

As for von Neumann, we can evoke what he explained in his last legacy, written a few weeks before his death, a moving book written with urgency, which he threw himself into with every last ounce of strength, *The Computer and the Brain*[46]. He declares in the final paragraph, entitled "The Language of the Brain, Not the Language of Mathematics" that:

> Thus logics and mathematics in the central nervous system, when viewed as languages, must structurally be essentially different from those languages to which our common experience refers.

---

45 Turing, A. (1954). Solvable and Unsolvable Problems. *Penguin Science News*, 31, 23.

46 Published in paperback, Princeton University Press.

What more can we say? Intuition does not come from modeling as Gödel and Turing clearly established. We must create the grammar of this multi-faceted new language. In order to navigate the infinitely complex while controlling the risks incurred, we must proceed step by step, without trying to skip any stages. A well-mastered grammar, the dual language/metalanguage approach, is an open structure that allows us to make sentences that can have one sense for an external user while preserving the general internal logic of everything that has already been validated; here again, the metaphor of the puzzle, introduced by Turing, maintains its relevance, for in the end, language is what allows us to add one more piece to the "meaning map" that we are building. Between the two, there is an interface of transduction that follows the integrative logic of "more is different" with a margin of tolerance that accounts for the specificities of each level[47]. Von Neumann's last sentence, which could appear sibylline, contains the perfect illustration of the system stack in the smartphone, if we are happy to consider the languages that are at both extremes of the chain. On the one hand, we have the programming languages that the user can see and activate using his thumbs on the touch screen[48]. But at the other extreme, we have the language of the masks piloting the silicon engraving machines, on the basis of let us say VHDL, for which we concretely create the nanometric energetic transducers that will in the end do the work. There is no relationship between them, structurally speaking, except that they say the same thing.

Biological information, as Dessalles and his colleagues define it in their book, *Le fil de la vie*, suggests that if there is a language, there is necessarily a grammar, as there is a grammar of nature, for the moment still restricted to hierarchy level H1, although we do not yet know the rules, and we only see traces of it in hierarchy H2. But perhaps hierarchy H3, supported by H1, will give us new ideas, and as Feynman would say, we will "stick our neck out" (Figure 6.5). It is more than probable that our brain, and neurons in general, have the means to "calculate" by making use of a process based on chance, which is made possible by the widespread parallelism used by ultra-specialized cells that appeared very early in the history of evolution.

Von Neumann, Wiener, Shannon and Turing were surely the first to have fully understood the almost "magical" virtues of chance.

Von Neumann's contribution regarding the approximation methods of nonlinear problems, especially in what is known as the Monte Carlo method, based on an approach of random sampling, was a complete turning point. He was very familiar with Borel's work, given that the game theory he developed in the 1940s was based

47 At this stage, we could cite PAC logic (Probably Approximately Correct), described by Leslie Valiant, Turing Award 2010, in the book with the same name that was recommended to us by Kourilsky.

48 See section 6.4.2.

on an initial study by the latter. Robotic theory played a central role in it since it helped to organize the appearance of events that could thus be controlled. This was the surest method for implementing a probabilistic PAC logic; he explained this in the five lectures on *Theory and Organization of Complicated Automata*, found in *Part 1* of the posthumous volume, *Theory of Self-Reproducing Automata*, edited by Burkes in 1965. The crucial point of the method concerns how random numbers are obtained, hence his advice, related by his biographers:

> Anyone who considers arithmetical methods of producing random digits is, of course, in a state of sin. For, as has been pointed out several times, there is no such thing as a random number – there are only methods to produce random numbers, and a strict arithmetic procedure of course is not such a method.

As we have recalled throughout the second part of this book, every rational number is, by definition, a periodic number whose period can be large, but always finite. Which means that any arithmetical method, and moreover, any computer-based calculation, mechanically introduces a statistical bias. We only have, in fact, pseudo-random numbers, which means that this interesting property disappears progressively[49]. We must therefore be extremely careful and be aware of the number of calculation steps needed, in other words, the logical depth, so we do not delude ourselves.

This point was especially misunderstood by the proponents of chance and noise – criticized by Thom – and in fact, the issue is absent from the criticized works, or sometimes evoked but with overwhelming contradictions that defy basic logic. It is when we are faced with the structure or real systems like the ones developed by engineers – von Neumann, Shannon and Turing were also engineers, and they demonstrated this – that we can re-establish the truth of the matter. It is thus never too late to do the right thing.

6.4. Testimonials: another look at the interface stack

6.4.1. *Harthong, Feynman, von Neumann and Turing*

As previously indicated, the physicist Richard Feynman – a 1965 Nobel laureate, discoverer of quantum electrodynamics and, incidentally, the person who posited the foundational thoughts that led to quantum information, taken up by Wheeler, who was his thesis director – constantly brought our attention to the application of mathematical formalisms, including probabilities, for the relationship they have with

49 Some equipment used in cryptography makes use of quantum phenomena to create random numbers that, because of this, escape the bias of arithmetic calculations.

reality[50]. The text that follows, cited by Harthong, comes from the *Feynman lectures* (Volume 3, *Quantum Mechanics*, Chapters 2–6 "Philosophical implications")[51].

NOTE 6.4.– We should note that all these texts predate, by some years, the publication of the book by Jacques Monod, *Chance and Necessity*, in 1970. The way that physicists speak of chance has nothing to do with the way biologists do, even very recently. It would be great to work on putting the different points of view into dialogue[52]!

Harthong says:

> There are problems posed by physics for which *no one* initially knows the role of pure chance, and finding it is a job for a genius. The most spectacular historical example is obviously the discovery of quantum mechanics: starting with experimental results […] the level at which pure chance played a role had to be found through research; we still have not found it. It would therefore be absolutely naive to attempt to develop a general and systematic method for finding it[53] (p. 21).

A little later, he adds:

> The question of the determinism or indeterminism of Nature is simply devoid of any logical sense. For this question to have one, we would need to have a rigorously exact and non-approximate mathematical understanding. The existence of innumerable philosophical explanations on the subject is not enough to transform this absurd question into a reasonable question (p. 27).

This is where he leaves it to Feynman to finish the thought. Feynman, responding as the great physicist he was, gives his lesson[54].

---

50 Already cited.

51 See: http://www.feynmanlectures.caltech.edu/III_toc.html – the lecture from 1965.

52 This appears to be an old debate. In a book from the 1950s, before the discovery of DNA, *Problématique de l'évolution* by Professor François Meyer, the problem of chance was posed in the same terms as this study. The book refers to other, even older works, all of which incidentally end up at Borel's book, *Le hasard*, already cited; see Alexys Moyse, *Biologie et physico-chimie*, P.U.F., and Guye, C.-E. (1922). *L'évolution physico-chimique* (cited by Léon Brillouin in his article "Life, Thermodynamics, and Cybernetics", 1949).

53 This fundamentally contradicts the pretense of gurus and prophets of Big Data who see in it an avalanche of data and AI, for example, Bostrom et al., substituting well-constructed theoretical knowledge, with the hypotheticodeductive method used sensibly, as with Turing, von Neumann, Grothendieck, etc.

54 The underlined passages have been emphasized by the author of this book to signal the importance of the statement.

Let us consider briefly some philosophical implications of quantum mechanics. As always, there are two aspects of the problem: one is the philosophical implication for physics, and the other is the extrapolation of philosophical matters to other fields. When philosophical ideas associated with science are dragged into another field, they are usually completely distorted. Therefore we shall confine our remarks as much as possible to physics itself.

First of all, the most interesting aspect is the idea of the uncertainty principle; making an observation affects the phenomenon. It has always been known that making observations affects a phenomenon, but the point is that the effect cannot be disregarded or minimized or decreased arbitrarily by rearranging the apparatus. When we look for a certain phenomenon we cannot help but disturb it in a certain minimum way, and the disturbance is necessary for the consistency of the viewpoint. The observer was sometimes important in prequantum physics, but only in a trivial sense. The problem has been raised: if a tree falls in a forest and there is nobody there to hear it, does it make a noise? A real tree falling in a real forest makes a sound, of course, even if nobody is there. Even if no one is present to hear it, there are other traces left. The sound will shake some leaves, and if we were careful enough we might find somewhere that some thorn had rubbed against a leaf and made a tiny scratch that could not be explained unless we assumed the leaf were vibrating. So in a certain sense we would have to admit that there is a sound made. We might ask: was there a sensation of sound? No, sensations have to do, presumably, with consciousness. And whether ants are conscious and whether there were ants in the forest, or whether the tree was conscious, we do not know. Let us leave the problem in that form.

Another thing that people have emphasized since quantum mechanics was developed is the idea that we should not speak about those things which we cannot measure. (Actually relativity theory also said this.) Unless a thing can be defined by measurement, it has no place in a theory. And since an accurate value of the momentum of a localized particle cannot be defined by measurement it therefore has no place in the theory. The idea that this is what was the matter with classical theory is a false position. It is a careless analysis of the situation. Just because we cannot measure position and momentum precisely does not a priori mean that we cannot talk about them. It only means that we need not talk about them. The situation in the sciences is this: A concept or an idea which cannot be measured or cannot be referred directly to experiment may or may not be useful. It need not exist in a theory. In other words, suppose we compare the classical theory of the world with the quantum theory of the world, and suppose that it is true experimentally that we can measure position and momentum only imprecisely. The question is whether the ideas of the exact position of a particle and the exact momentum of a particle are valid or not. The classical theory admits the ideas; the quantum theory does not. This does not in itself mean that classical physics is wrong. When the new quantum mechanics was discovered, the classical people – which included everybody except Heisenberg, Schrödinger, and Born – said: "Look, your theory is not any good because you cannot answer certain questions like: what is the exact position of a particle?, which hole does it go through?, and some others." Heisenberg's answer was: "I do not need to answer such questions because you cannot ask such a question experimentally." It is that we do not have to. Consider two theories (a) and (b); (a) contains an idea that cannot be checked directly but which is used in the analysis, and the other, (b), does not contain the idea.

If they disagree in their predictions, one could not claim that (b) is false because it cannot explain this idea that is in (a), because that idea is one of the things that cannot be checked directly. It is always good to know which ideas cannot be checked directly, but it is not necessary to remove them all. It is not true that we can pursue science completely by using only those concepts which are directly subject to experiment[55].

In quantum mechanics itself there is a probability amplitude, there is a potential, and there are many constructs that we cannot measure directly. The basis of a science is its ability to predict. To predict means to tell what will happen in an experiment that has never been done. How can we do that? By assuming that we know what is there, independent of the experiment. We must extrapolate the experiments to a region where they have not been done. We must take our concepts and extend them to places where they have not yet been checked. If we do not do that, we have no prediction. So it was perfectly sensible for the classical physicists to go happily along and suppose that the position – which obviously means something for a baseball – meant something also for an electron. It was not stupidity. It was a sensible procedure. Today we say that the law of relativity is supposed to be

55 A fundamental point. In the atomic sphere, by definition, one "sees" nothing. Atoms became an indispensable logical necessity for theoretical coherence before they could be observed. With this, we are at the foundations of the hypotheticodeductive method. The consequences of the atomic hypothesis are, themselves, easily observable.

true at all energies, but someday somebody may come along and say how stupid we were. We do not know where we are "stupid" until we "stick our neck out", and so the whole idea is to put our neck out. And the only way to find out that we are wrong is to find out what our predictions are. It is absolutely necessary to make constructs[56].

We have already made a few remarks about the indeterminacy of quantum mechanics. That is, that we are unable now to predict what will happen in physics in a given physical circumstance which is arranged as carefully as possible. If we have an atom that is in an excited state and so is going to emit a photon, we cannot say when it will emit the photon. It has a certain amplitude to emit the photon at any time, and we can predict only a probability for emission; we cannot predict the future exactly. This has given rise to all kinds of nonsense and questions on the meaning of freedom of will, and of the idea that the world is uncertain.

Of course we must emphasize that classical physics is also indeterminate, in a sense. It is usually thought that this indeterminacy, that we cannot predict the future, is an important quantum-mechanical thing, and this is said to explain the behavior of the mind, feelings of free will, etc. But if the world were classical – if the laws of mechanics were classical – it is not quite obvious that the mind would not feel more or less the same. It is true classically that if we knew the position and the velocity of every particle in the world, or in a box of gas, we could predict exactly what would happen. And therefore the classical world is deterministic. Suppose, however, that we have a finite accuracy and do not know exactly where just one atom is, say to one part in a billion. Then as it goes along it hits another atom, and because we did not know the position better than to one part in a billion, we find an even larger error in the position after the collision. And that is amplified, of course, in the next collision, so that if we start with only a tiny error it rapidly magnifies to a very great uncertainty. To give an example: if water falls over a dam, it splashes. If we stand nearby, every now and then a drop will land on our nose. This appears to be completely random, yet such a behavior would be predicted by purely classical laws. The exact position of all the drops

56 The idea of "construction", in the sense of engineering/modeling, is central to Feynman's work. On his blackboard, in his office at Caltech, towards the end of his life, his visitors could read the injunction "what I can't create, I can't understand" (source: Caltech archives). This approach is founded on constructive mathematics, so important to Poincaré, who lived through a truly revolutionary period.

> depends upon the precise wigglings of the water before it goes over the dam. How? The tiniest irregularities are magnified in falling, so that we get complete randomness. Obviously, we cannot really predict the position of the drops unless we know the motion of the water absolutely exactly.
>
> Speaking more precisely, given an arbitrary accuracy, no matter how precise, one can find a time long enough that we cannot make predictions valid for that long a time.
>
> Now the point is that this length of time is not very large. It is not that the time is millions of years if the accuracy is one part in a billion. The time goes, in fact, only logarithmically with the error, and it turns out that in only a very, very tiny time we lose all our information. If the accuracy is taken to be one part in billions and billions and billions – no matter how many billions we wish, provided we do stop somewhere – then we can find a time less than the time it took to state the accuracy – after which we can no longer predict what is going to happen! It is therefore not fair to say that from the apparent freedom and indeterminacy of the human mind, we should have realized that classical "deterministic" physics could not ever hope to understand it, and to welcome quantum mechanics as a release from a "completely mechanistic" universe. For already in classical mechanics there was indeterminability from a practical point of view.

6.4.2. *Internal language, external language, according to von Neumann*

In the final message from von Neumann, recalled in section 6.3, we pointed out that the information stack perfectly illustrates this dichotomy of languages, with an external language on one side, oriented towards users, and the internal language, organically oriented with its energy transducers, the former being compiled/translated and/or interpreted within the latter. Here, we will provide some irrefutable complements to this integrative hierarchical vision of "more is different".

This distinction is essential if we are to properly understand the nuance and/or articulation of the real and the virtual. We will illustrate this for the cases where drone-based combat systems along with many other kinds are used; but we could likewise take other types of systems of systems (SoS) – this is the name habitually given to them – like the production/distribution of energy using smart grids, smart cities or even trading systems with their high-frequency trading (HFT) indices[57], air traffic control, etc.

57 See: https://en.wikipedia.org/wiki/High-frequency_trading; legal pillaging done intelligently!

In every scenario, the problem is the same, at least formally speaking; in other words, building a virtual image that can be manipulated through computational means, an image which is connected to the real world through a set of sensors/receivers, in great quantity, tied to computing systems that make up the SoS, with everything under the control of the users.

We therefore have the first retroactive look at the top of the stack, between the real world and the computerized virtual world – a COP or a Common Operational Picture (Figure 6.9).

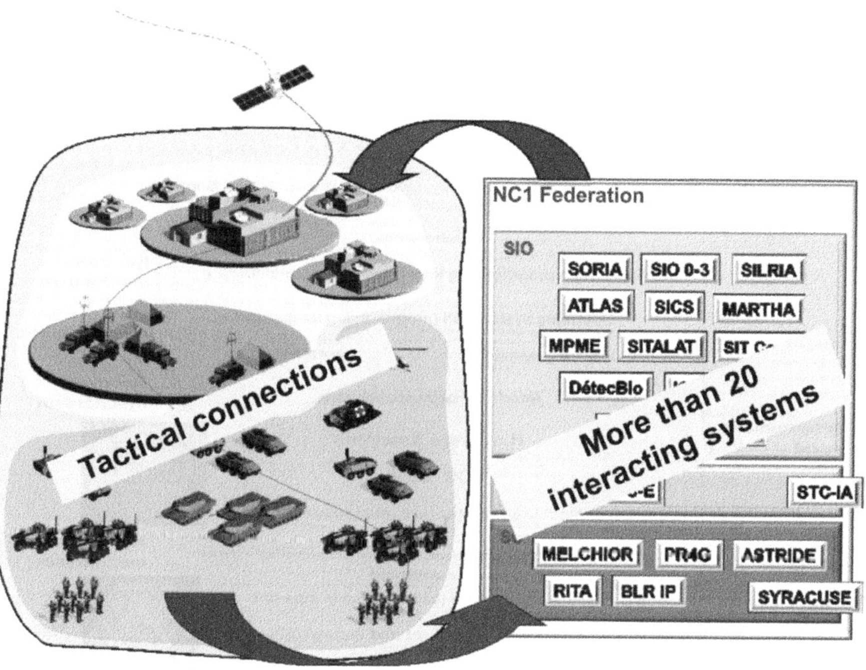

Figure 6.9. *Real world/virtual world correspondences*

The user of the system "sees" the real situation through an image that is expressed using a graphic language that is specific to this type of system, which all users must master in order to coordinate and give meaning to their actions. This language constitutes the most external interface of the overall SoS system[58]. It is thanks to this language that users are able to act and coordinate. But for the action to be efficient, it is essential that the virtual, abstract image be as exact as possible,

58 See the reference document: http://www.dtic.mil/doctrine/doctrine/other/ms_2525c.pdf; *Common Warfighting Symbology*.

given the specificities of the equipment and the actors "on the field," various delays and errors of every kind that can randomly appear, including *fakes*, which are also part of the issue; given the quantum nature of the energetic interactions and the time for them to propagate, there will always be a delay, hence the importance of the PAC logic[59]. The operator only sees an approximation.

From there, any actions taken by the operators/users will be translated into a succession of steps, as can be seen in Figure 6.10; this will continue until the energetic transducers that in the end make the selected function happen are reached, which will make it possible to act in the real world using means of "actuators," some of which will be human actors who will be informed of what they must do via text, voice and image messages (the case of computerized foot soldiers in the French FELIN system or the American Future Combat Systems)[60].

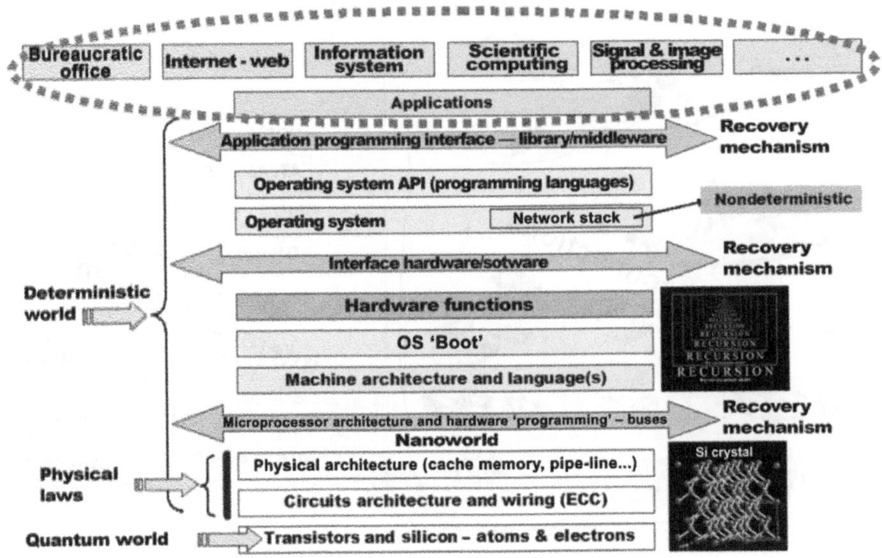

Figure 6.10. *Architecture in layers*

6.4.3. *Imbrication of stack languages*

Returning to von Neumann's thesis, the question is therefore how we compare the languages we encounter in this recursive descent, with the difference that here we are entirely in control of what we do, which is not the case for the language of

59 *Common warfighting symbology*, already cited.

60 See: https://en.wikipedia.org/wiki/F%C3%89LIN.

the brain, which we do not actually know to exist, although we can make a reasonable hypothesis that it necessarily must.

We can have, in a stack, "observers" who will allow us to understand what is happening in it, as indicated in Figure 6.11. These observers are constitutive of the monitoring system, which is indispensable for the validation of the system.

The initial action of the user (the initial state) will be translated by a certain number of processes that will be executed by a structure like the one in Figure 6.7, which means that starting with this action, tens/hundreds/thousands of processes will be initialized and the final overall result will produce the desired action (the final state), which in turn will be evaluated in order to determine the next steps, etc. The combinatorial of intermediate states created by this parallel execution is an inaccessible number.

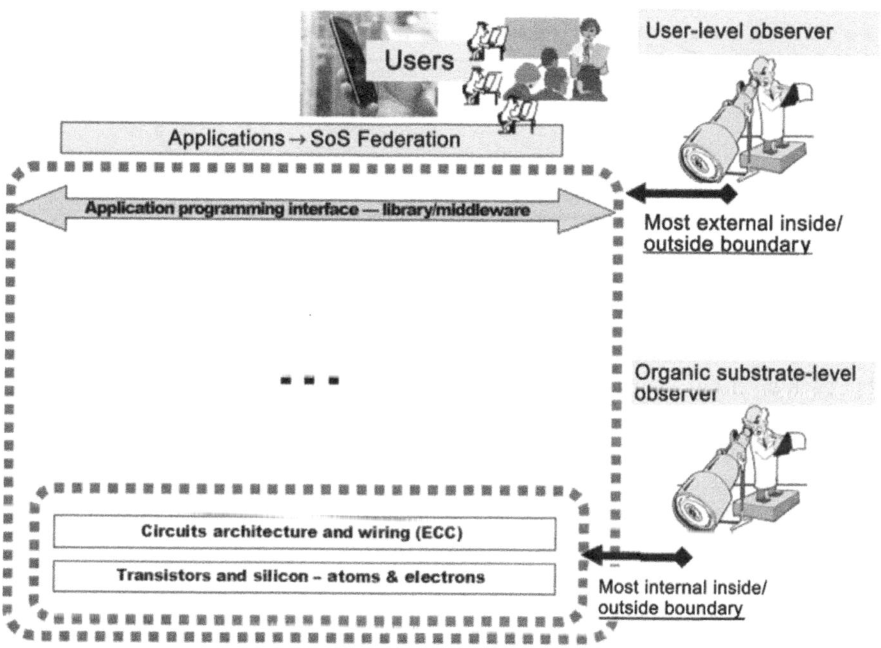

Figure 6.11. *Hierarchy of observers*

This is therefore the question: can we understand the intentions of the user by observing what happens at an organic level? Or, in a less restrictive way, at which level of the stack does the connection between the user's intention and what happens in the level being considered cease to be comprehensible? This is a problem of retro-engineering such as the ones engineers must solve each time there is a manifest error whose source must be found in order to correct the cause. In other words, is the transformation that has been made reversible? If it is not, can we make it reversible, and if so, at what energetic cost?

To understand the problem of reversibility, we must examine the nature of the translation/transduction made by looking very precisely at what happens at the interface (Figure 6.12).

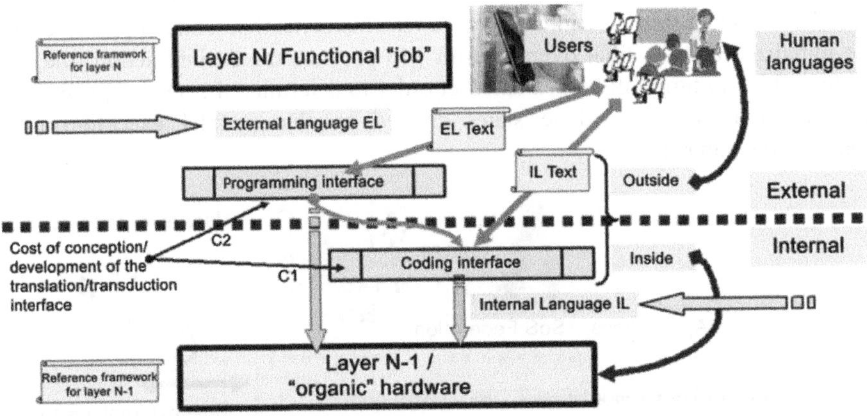

Figure 6.12. *Translation/transduction interface*

To avoid overloading the figure, only the users of layer *N* have been represented, but each layer has its users who will act according to the specific rules of the layer, rules that are specified in the reference framework for the layer. Each layer therefore has its own "language", its grammar, and the users of the layer express themselves in the language of the layer. For translation/transduction to happen, users of both interacting layers must share a certain number of conventions that make it possible to translate between the external language (EL) and the internal language (IL); by doing this, we establish an abstract correspondence EL → IL that guarantees the semantic equivalence of the two texts. If the execution of the translated text does not produce the expected result, we must be able to go back to the initial text in the EL, correct it and/or change the translations rules presented in the reference framework. Therein lies the property of reversibility.

We should note, in passing, that this manner of working strictly adheres to the general principles of more is different, and that incidentally, the rules of translation are compromises judged to be acceptable to the users (PAC logic, which they are the only ones to master).

For example, a translation rule can establish a threshold of tolerance concerning the duration of the operation, which will in turn allow user N to act while respecting the rules that apply to them but that user $N-1$ has no reason to know. The user of a computer must obviously know the rules of the languages they use, perhaps up to the level of the computer's "assembly" language, but they have no reason to know the ones that pilot the energetic transformations at the level of the computer's transducers, and even less the ones that pilot the transformation of electrical signals being exchanged between components. In the mathematical language of signal processing, these are represented by "distributions", that is, sets of functions.

This principle of voluntary but controlled "ignorance" has a fundamental character since it is what enables the ascent towards abstraction, which characterizes the integrative logic of the stack, a principle that lets user N "forget" the internal logic of layer $N-1$, all the while trusting the users of this layer and the agreements they have made with them.

In doing this, the user preserves the constant that corresponds to their capacity to control – a specific property – which is a fundamental systemic principle; the price to pay is absolute trust that users have in the two interacting layers.

In an operation such as $a+b=c$, if c equals, let us say 5, and this causes an error, and we have lost the values of a and b, we will have to try all the combinations of values $\{a, b\}$ such that $a+b=5$, which can lead to numerous attempts, given the information we may have about variables a and b (e.g. 32-bit binary digits but with characteristics that restrict the possible values) or clues that we might find in the execution context. It is also necessary to find the trace of any references to a and b, not only in the program containing the instruction $a+b=c$, but also in other programs where a and b might appear under different names, given the rules of the programming language of the layer. This is the logical problem, known for some time, of the meaning and the notation of entities, omnipresent in programming[61] Church, in his *Introduction to Mathematical Logic*, gave the example of the planet known as Venus, which has two meanings: (1) morning star and (2) evening star, depending on the period of the year when it is being observed.

Any computer connected to the Internet has an IP address (e.g. the number 172.16.254.1 in the IPv4 norm) that is the notation for the computer, its unique name in the network, but which is never used as such by external users[62]. The users are more likely to use its URL, which is a given name, that is, a meaning that identifies accessible resources on the Internet, like, for example, what identifies a Wikipedia

61 This notion was introduced by Gottlob Frege; https://en.wikipedia.org/wiki/Sense_and_reference.

62 See: https://en.wikipedia.org/wiki/IP_address.

document explaining what a URL is: https://en.wikipedia.org/wiki/URL. We can incidentally find it by providing the search criteria to our favorite search engine, which is another way of naming things, but which can end up being dangerous if there are many responses/possible targets.

In a general way, all created computational entities must receive a unique notation, its unique name, but will be able to be referred to in different ways depending on the contexts of use, which will facilitate the management of programs and allow for their progressive evolution.

The moral of the story is, if we want to recreate the steps that led to an erroneous value of *c*, we must follow back all the references to *a* and *b*, under all the names by which the entities *a* and *b* are known, by combining the possible authorized values. We must also find all the programs in which *a* and *b* appear under their different names, programs that can simply be partially erroneous, which might have led to the erroneous value of *c* noted in the first place. Since the programs in question can be executed in different, parallel flows, as in Figure 6.7, we must also find the information that caused synchronization to be done in such or such a way[63].

This type of problem is very familiar to the architects of databases since the proper management of names saved in the database is what guarantees its integrity[64]. Randomly repeating data in a database is, ipso facto, creating the potential for an ambiguity, which, associated with the mechanisms for naming, meaning/notation, will make reversibility impracticable, for one duplicate piece of information is worth two notations for an entity that is a priori unique, which will in turn progressively create an incontrollable combinatorial whose quick propagation – usually exponential – will always surprise architects.

The apparently trivial problem of finding what may very well have happened to explain the erroneous value of *c* that was noted is, in fact, a formidable combinatorial problem which, for the same reasons as those evoked in section 5.5.1, will create for us these inaccessible numbers that engineers who know what they are doing fear dearly. And so forth, for each level of the stack where there is programming, that is, up to the very heart of the SOC circuits up which everyone

63 In technical terms, this boils down to tracing the product of robots that communicate, whose theory was created by my two colleagues, André Arnold and Maurice Nivat, among others; all these problems are of a non-polynomial complexity, and therefore generators of inaccessible numbers.

64 Non-reproducible errors are the ones that are feared the most; some authors such as Jim Gray qualify them as *Bohrbugs* or *Heisenbugs*, depending on the case, in reference to the physicists Niels Bohr and Werner Heisenberg, since it was enough to want to observe them for them to disappear.

engaged in intense programming is embarking, with the micro-nodes and external drivers integrated in the SOC, programming that reaches up to millions of lines of code.

In the practice of complex artificial systems engineering, reversibility is obviously always possible since if it were not the case, we would quit what is commonly called engineering, where we know what we are doing given the know-how, and instead enter the world of magic or the occult sciences. It is even required in critical systems[65]. This is also the reason why we constantly monitor systems, to ensure that the hypotheses that prevailed at the time of their manufacture, what we call the reference framework of the system (the nominal state of the system), are still respected.

On the contrary, if the reference frameworks have disappeared or are hidden for reasons of confidentiality and/or the protection of intellectual property, climbing from the bottom to the top of the stack is strictly impossible for the simple and good reason that there are interfaces we have precisely designed so that the external user would not have to know what happens inside, except for what was agreed upon among the respective parties, such as, for example, the return codes that provide information of a qualitative nature (done, not done, waiting, etc.) and allowing the external user to continue their work. Wanting to go beyond that means encountering a combinatorial problem that can become uncontrollable because of inaccessible numbers. At each stage of the computer stack, the logic of common sense prevails, as has already been said, for at this level, everything is a matter of conventions between users. What is seen as information/action by an external user is divided into thousands of micro processes at the organic level, potentially spread over hundreds or thousands of computers that synchronize themselves however they can, depending on the rules of synchronization collectively established by programmers, but whose dynamic itself remains unknown[66]. This is a gigantic puzzle for which we know neither the semantic partitions nor the assembly rules, a puzzle in which semantics are irredeemably linked to the rules of construction – syntax and pragmatics, the modalities for correction/compensation of errors, etc. – specific to each layer. The – several hundred – instructions available on a computer may certainly be finite in number, but the way of arranging them to respond to a specification is an inaccessible number, which guarantees that if we entrust the programming to two programmers who do not communicate with each other, we will get two different programs, an interesting property that is taken advantage of in *N*-version programming since these two programs can never produce an error in the same way. We thus see why puzzles so fascinated Turing, the man who solved the

65 In nuclear energy, in avionics, and probably very soon for self-driving cars; see the DO-178B norm, https://en.wikipedia.org/wiki/DO-178B.

66 See the monograph by Dijkstra, E., *Cooperating sequential processes*, downloadable.

most famous one, that of the Enigma machine, a machine he knew to have been invented by men to transmit operation orders to equally known entities whose location was unknown, but whose real existence was never in doubt[67].

We can thus consider that since the arrival of VLSI/SOC such as the Pentium IV from Intel, with its 55 million transistors[68], the reverse engineering of a circuit has become practically impossible, on the one hand, because of its small size, and on the other hand, because of the number of layers needed for the connections between components, currently up to 10–15 layers, making these connections entirely inextricable in the literal sense of the term. This also makes it possible to hide something that we do not want found inside. Betrayal is always possible.

It goes without saying that it is entirely impossible to understand the functionality of a complex circuit such as the one in the smartphone, in Figure 6.1, by manipulating it from the outside, very simply because this is not the object of the average user's concerns, and also because the designers have taken every precaution to ensure that the user will not have to worry about this, and from this angle, it is a perfect success.

Another equally fundamental property of the interface architecture depicted in Figure 6.12 concerns its evolution. Artificial systems evolve because the needs of users evolve, and because the technology also evolves. The system must therefore constantly be adapted so that it can remain optimized for survival, an operation which is of the same nature as the correction of an error. The most powerful mechanism that engineers have found for resolving this difficult problem also relies on the EL/IL language coupling. Any information about layer $N-1$ that enters layer N at the wrong moment would have the consequence of coupling the two layers; the one could no longer evolve without the other, which means that the entire edifice would be destabilized during a modification. This is compensated for by exploiting the language/grammar pairing and the relative stability of the grammar with respect to the phrases it can accept. To enrich an interface is to first enrich its language (the vocabulary, for example) by leaving the grammar the same; but we can also enrich the grammar, which is a more delicate operation, since this enrichment must not create any incoherence. The addition of new rules should therefore not create any

---

67 His article, "Solvable and unsolvable problems", already cited, is entirely based on the metaphor of the puzzle. He begins with these words: "If one is given a puzzle to solve one will usually, if it proves to be difficult, ask the owner whether it can be done".

68 To truly understand the significance of this number, we can compare it to that of the Motorola 68000 chip from the 1980s that contained 68,000 transistors for a simplified but complete CPU. The computer that I worked on the most had a very complete CPU of about 300,000 transistors; that means that the circuit of the Pentium IV contains many other things besides a CPU, beginning with cache memory.

incompatibility. Hence, the principle of upward compatibility and its equally important variant: the bootstrap, which is when a program makes another program using a description of the latter, as with metacompilers (or compilers of compilers, in the jargon), which have been in use since the 1980s.

Logically speaking, this comes down to enriching the axioms of a theory in order to diversify it without putting the initial theory at risk, as illustrated by David Hilbert with his logical model in his famous treaty, *The Foundations of Geometry*[69].

Depending on the case and the execution constraints, the text presented to the interface will be compiled and/or interpreted, but this is another story that relates to computational "cooking". The most famous protocol on the Web and the W3C is the http protocol, which billions of Internet users trust; it was built according to this principle[70].

The extent to which all of this would be applicable or transposable in the living world remains an open question, one to which systems architects and programmers can obviously not provide a response. The metaphor of the computer stack nevertheless shows that a knowledge of organizational levels and the barriers of specificity – in the sense that Kourilsky gives this expression – which make up the material interfaces is indispensable if we are to understand the phenomena of emergence/integration, *from the bottom down* as the Nobel laureate Robert Laughlin would say. Without these hierarchical levels, without the reference frameworks that are associated with them, an object like the smartphone becomes entirely incomprehensible. It seems obvious, however, that the human brain, via human languages, which are moreover all translatable into each other and are purely circumstantial, as von Neumann pointed out, materializes a capacity for abstraction[71]. This capacity implies a radical separation of the functional aspects of organic matter via the common memory, where data and instructions end up being represented by this very same code, binary {0,1} in our computers, but interpreted differently by the CPU; or even that which allows us to connect what is discreet/countable in language with the geometric continuum of our visual perception, as with Bézier's curves. This is what von Neumann had already remarked, and what fascinated him so much.

---

69 Printed in German: *Grundlagen der Geometrie*, Leipzig, Teubner, 1899, translated into English in 1902; constantly reedited. See: https://en.wikipedia.org/wiki/Hilbert%27s_axioms.

70 See: https://en.wikipedia.org/wiki/Hypertext_Transfer_Protocol; and the standard, with its eight basic commands, https://www.w3.org/Protocols/rfc2616/rfc2616.html.

71 *The Computer and the Brain*, pp. 80–81. Also see the book by Piaget, J., already cited, *L'équilibration des structures cognitives*, P.U.F.

The place where these interfaces reside within a conscious, living system, as may be seen in the development of the hominid lineage, will remain completely mysterious for the time being, but this is not because we do not see them or because they do not exist. They are logically necessary, as were atoms, well before we could actually "see" them.

PART 3

Examples of Systemics and Complexity

Introduction to Part 3

In this third and final part, we aim to illustrate some of the principal aspects of the fundamental concepts of systemics summarized in Chapter 1, by way of examples.

For this, we have chosen two emblematic examples of complexity: (1) the French electrical system, one of the largest systems ever made in the world, which dates back to the start of the 20th century and which is used by everyone in France and (2) the organization of engineering teams who deliver systems such as the ones discussed in this book.

The electrical system is a very large technological system producing and distributing a colossal amount of energy and one that few people are aware of. For it to function properly, not only is it necessary to have a perfect understanding of the electrical phenomena discovered over the course of the 19th century, but also of the information needed for the integration and control of thousands of pieces of equipment. The engineers working on it were and are still trained accordingly, including nuclear technology from the 1960s onwards. Looking at how it is done, how certain problems have been resolved, in particular those posed by renewable energy, which disrupts the balance of offer and demand, the need to never lose sight of the three laws of systems becomes easier to understand[1].

The organization of engineering teams, and this goes well beyond the scope of engineering personnel, into what we generally call "projects", is essentially a social science that consists of harmonizing everyone's individual efforts into a coherent whole with the help of an integrated social dynamic. Developing trust, and trusting ourselves: this is the challenge of the science of projects. In a system whose creation can include the participation of thousands of engineers and technicians, it is

1 See *System Architecture and Complexity*, already cited.

impossible for one single person, and even for one small team, to materially control everything that happens and to have every aspect report back to a single, decision-making central office. As we sometimes say, intelligence is "collective", but we still need to know what that means exactly. For a collective to work and reach its objectives, it must give itself rules, and these rules must be understood in order to be shared and applied by everyone. As we will see, these rules constitute a reference framework that is the common property of the collective. If the actors scrupulously apply the rules, then they will be able to develop mutual trust and their energy will accumulate. The reference framework is the contract that ties them together, and ethics, the moral quality of individuals, become a primordial parameter. The problem is therefore not that people need to adapt to each other, but that they should adapt to the reference framework until the moment where adapting the reference framework itself becomes necessary. If the actors do not trust each other, that is, if they do not trust the reference framework that is their common property, then cooperation becomes impossible. The capacity for organization becomes greatly limited, as does the complexity itself. The project's reference framework operates somewhat like a language that is not the property of anyone but for which all interested parties are guarantors.

Here again, we will see that systems give an overarching view of these human processes, without which none of these systems that make our lives so easy would have been able to appear and be developed. It is the key to collective action, well beyond the things that engineering is stricto sensu concerned with; it is a cultural object, perhaps even a civilizational object, a distant consequence of the "demonstrative order" that emerged in the Greek world during the 7th and 8th centuries BCE and upon which our rationality is built.

7

Systemic Aspects of the French Electrical System

ACKNOWLEDGMENTS.– Writing this chapter would not have been possible without the contributions from Mr. Yves Bamberger, the advisor to the Chairman at EDF and a member of the Academy of Technology, and Mr. Pierre Bornard, Vice President of RTE and President of PowerNext during the time when the cycle of conferences was being organized by the CESAMES association[1]. We thank them, in particular, for all their time, for the recommendations they generously gave us and for making images available for our use.

The content of this chapter naturally only represents the perspective of the author.

The electrical system (ES) – production, transport and distribution of electrical energy to individuals and to companies – is one of the most complex technical objects, in the sense used by Simondon, that we are able to observe and study, even in its most minute details, since it is entirely created by human intelligence[2]. Simondon conveyed the central message of systems engineering when he explained that a technical object is inseparable from two human communities: the one that invents, makes and manages it, and the one that uses it. Engineering and R&D, on

For a color version of all the figures in this chapter, see www.iste.co.uk/printz/pedagogy.zip.

1 Text written following presentations by Y. Bamberger and P. Bornard at the CSDM 2013 conference, organized by CESAMES, and at the CESAMES workshops, *Intégration et Complexité*, in 2013.

2 See his thesis, *Du mode d'existence des objets techniques*, already cited. Also see: https://en.wikipedia.org/wiki/Gilbert_Simondon.

the one hand, and users, on the other hand, that is, approximately 20 million homes. Each of the communities has its own constraints and its own needs that must be harmonized to find a compromise that will satisfy both sides, with particular attention to the relationship between designers and operators, for the system must remain operational in all situations. At this level, we may speak of co-evolution.

The availability of abundant and high quality electrical energy is now so familiar to us that we tend to forget the brilliant minds and feats of engineering that needed to be rallied and deployed in order to satisfy the energy needs that our ancestors took centuries to conquer.

Figure 7.1. *Energetic mutation in the early 20th century*

We all still have images in mind of the first power plants where the energy produced by thermal machines, initially steam engines, was distributed along an axis that other machines would connect to in order to take what they needed using leather belts. The distribution constraints were those of the machine and its mechanical rigidity such that the human operator had no other choice but to adapt to it.

With electricity, this highly constrictive set up changed radically and underwent a mutation, a true metamorphosis, in the biological sense of the term. This was during the second wave of the industrial revolution at the turning point of the 19th and 20th centuries. Figure 7.2 provides a general blueprint of this.

The mutation came from the fact that the rigid axis, save a few gears that transferred mechanical energy from the source at the center of the factory, was replaced by a transport network made of flexible cables, electrical conductors and converters/transformers. In doing this, it became possible to arrange the machines more efficiently, leading to a better organization, more comfort, less noise, etc. The first production chains for cars, with Ford, were unimaginable without these kinds of configurations. It also became possible to replace gas-burning city lights, which were very dangerous because the gas could explode, with lamps like the ones we use today, or to bring electricity directly to the homes of individuals. Overall, it brought many advantages.

At the moment when the adventure of electricity began, electrical science was already well advanced thanks to genius experimenters such as Faraday, Edison and many others, alongside scientists Jean-Marie Ampère or James Clerk Maxwell, who published his famous treaty, *Treatise on Electricity and Magnetism*, in 1873. This was an important date for mathematical physics since, for the first time, theory predicted the existence of phenomena that would be discovered later, in this case the electromagnetic waves discovered by Heinrich Rudolf Hertz in 1890.

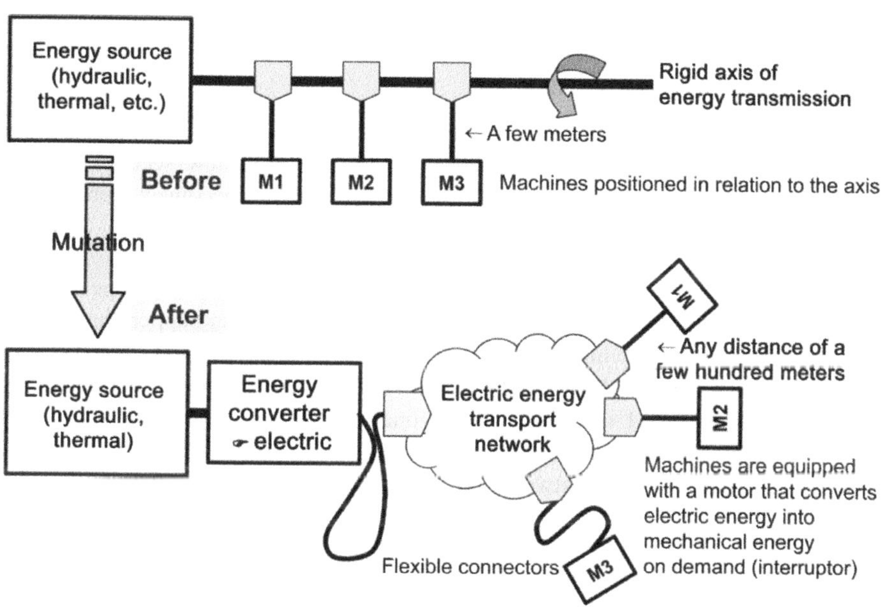

Figure 7.2. *Energetic mutation: overview*

Figure 7.2 also shows the basic element of any electrical system, an element formed of several devices: (1) an energy source, (2) a converter that transforms this energy into electric energy (dynamo, alternator), (3) a transport network for the electricity and (4) different kinds of equipment that reconvert the electrical energy as needed: light, heat, mechanical (electric motor) and even chemical as in the factories making aluminum through electrolysis.

This is a good example of active unity (AU), in the projects sense, which combines technical means and its two human communities (users *and* engineering in the general sense, conception *and* operations).

Nevertheless, this creates:

– an inconvenience: the electricity is not stored, or at least very badly stored (chemical batteries, cells, etc.). Whatever is produced must be consumed immediately, and conversely;

– two constraints:

- the electricity is easily transported, but at the price of energy dissipation (Ohm and Kirchhoff laws), so the distance between the locations of production and the locations of consumption must be optimized, otherwise the environment will get overheated and the return will diminish,

- the energy consumed must also never surpass the amount of energy produced by the machine, except for a small constant for inertia (a few seconds or even fractions of a second), otherwise the motor will stall and thus become a mere resistor waiting to melt; this is a blackout.

The notion of a system, in the systemic sense of the term, really appears when we attempt to combine geographically neighboring AUs. It can happen that AU1 is overproducing energy, while AU2 is not keeping up with demand. We therefore have *offer* AU1 > *demand* AU1 and *demand* AU2 > *offer* AU2. If we are able to set up a link between AU1 ↔ AU2, then the one can compensate for the other, such that the composition AU1 ⊗ AU2 satisfies the indispensable balance of *offer* = *demand*; without which the network short-circuits (a major incident that leaves users "in the dark").

This notion of compensation is at the heart of ES logic since we can see that establishing an exchange link makes it possible to offer users of AUs who cooperate a better service contract, even though the regulation link works perfectly well. There is a profound symmetry in this mechanism tied to the fundamental constant of system offer/demand.

We will not present the electrical system in detail, nor the relevant aspects of physics (several books would be needed for that) but we will recall some particularly significant points from the systemic approach, points that we find in all systems. Among other things, we should remember:

– ES growth in time and space.

– The percolation of the basic elements of the ES and the interoperability of these elements among themselves. How do we organize the cooperation of "electric" AUs and regulate exchanges for the benefit of users?

– The robustness/resilience of the ES when it is subjected to all sorts of hazards, whether internal and/or external, and in particular, how breakdowns are treated alongside the more or less predictable behavior of users[3]. How does it survive?

– The organization of the ES, in particular from the perspective of its progressive computerization. How do we improve the efficiency (the return) of the global system?

Many other points would be interesting, but that would be better suited to a book completely devoted to the systemics of these especially interesting types of systems, for which France is a world leader.

7.1. Growth: the "life" of an electrical system and its end

Any system satisfying the needs of its users will/can grow in two ways: it can (1) increase its number of users; and (2) increase the number of services offered, doing this until a position of equilibrium is reached where the "client" population is generally and durably satisfied. This growth in the number of users must happen without the service contract deteriorating and, if possible, by improving it.

The dynamics of this type of phenomenon are clearly modeled by systems of equations such as the Lotka–Volterra equations, which we find in many analogous phenomena in biology and/or kinetic chemistry, project management, etc[4]. One of these equations, which is used the most frequently, is that of the logistic function, more well-known as an S-curve because of its characteristic form: in geometry, this is the sigmoid whose differential form is $dA = k \times A(\varepsilon - A)dt$, with A designating a growth parameter that depends on the evolving parameter t (a form of "abstract" time), k and ε being the constants; the term $(\varepsilon - A)$ means that the linear growth slows down as the system develops.

3 See *Mémento de la sûreté du système électrique*, which can be easily downloaded from the RTE website.

4 Already cited, see Chapter 3.

Figure 7.3. *The French electrical system in the 1920s (source: Tribot-Laspierre, J. (1923). Les Lignes françaises de transmission d'énergie électrique à hautes et très hautes tensions. Le Génie civil, 26)*

REMARK 7.1.– If growth is constant, without slowing down, we end up with the classical exponential function. The derivative of the curve is a parabola whose peak corresponds to the inflection point of the curve. A discrete variant of this curve is the logistical series, also frequently used.

We can draw a first observation from this that is typical of systemics.

IMPORTANT.– To satisfy the need for growth, a community in charge of its engineering has two available solutions: (1) duplicate and/or reproduce the system identically, notwithstanding a few minor adaptations; or (2) augment the capacity/services of the constitutive elements of the current system, keeping in mind that beyond a certain size the element becomes unmanageable, and in this case must be duplicated.

For this, the community must have access to the system blueprint (a static model) and manage its evolution over time (a dynamic model), which requires a knowledge of the "grammar" rules that make it into a system.

In the first case, the extension of the system is spatial, in the space where the community of new users is found (see Figure 7.3); in the second case, the extension is temporal, with the system remaining locally stable all the while adapting its capacities as required.

Figure 7.3 represents the state of electrification in France in the 1920s. There were several companies producing electricity; the transport networks were partial and not connected among each other, and norms were almost inexistent.

We should note that the entire north and east as well as the valleys in mountainous regions were already well-provided for since that is where industrial activity was concentrated, especially the steel industry, mining and spinning mills, as well as the aluminum and electrochemical industries of the alpine valleys. EDF in its modern form was not created until 1945, after the war, through a merger of about 1,450 companies.

Figure 7.4 shows us how our (great)-grandparents were encouraged to consume even though it was still not quite the "electric fairy"[5].

Figure 7.5 shows us the current state of European networks (principally the north-east parts of France with its ties to Germany, the UK and Benelux, for the sake of legibility), now massively interconnected[6].

In a little less than a hundred or so years, thanks to the efforts of three to four generations of engineers, we have gone from a state where only the embryos of systems existed, all different and independent from each other, to a completely unified and synchronized system, providing energy to the most remote corners of the country, and capable of exchanges with the systems of neighboring countries.

In the interim, the electrical system became a vital necessity for the economy of the country, and there was no going back without some sort of cataclysm that would mark the end of social well-being and society as we know it.

5 An allusion to the painting by Raoul Dufy, the huge homage to electricity named *La Fée Electrique* ("The Electric Fairy"), commissioned by a Parisian electrical company for the 1937 Exposition Internationale.

6 For more details, see the ENTSO-E website: https://www.entsoe.eu/; a detailed map can be downloaded.

Figure 7.4. *Electricity advertisement from the 1920s. It outlines the ways of living, from the distant past, the past and today. It covers cooking, heating, washing, cleaning and watering but highlights that the best way to live well is to have electricity*

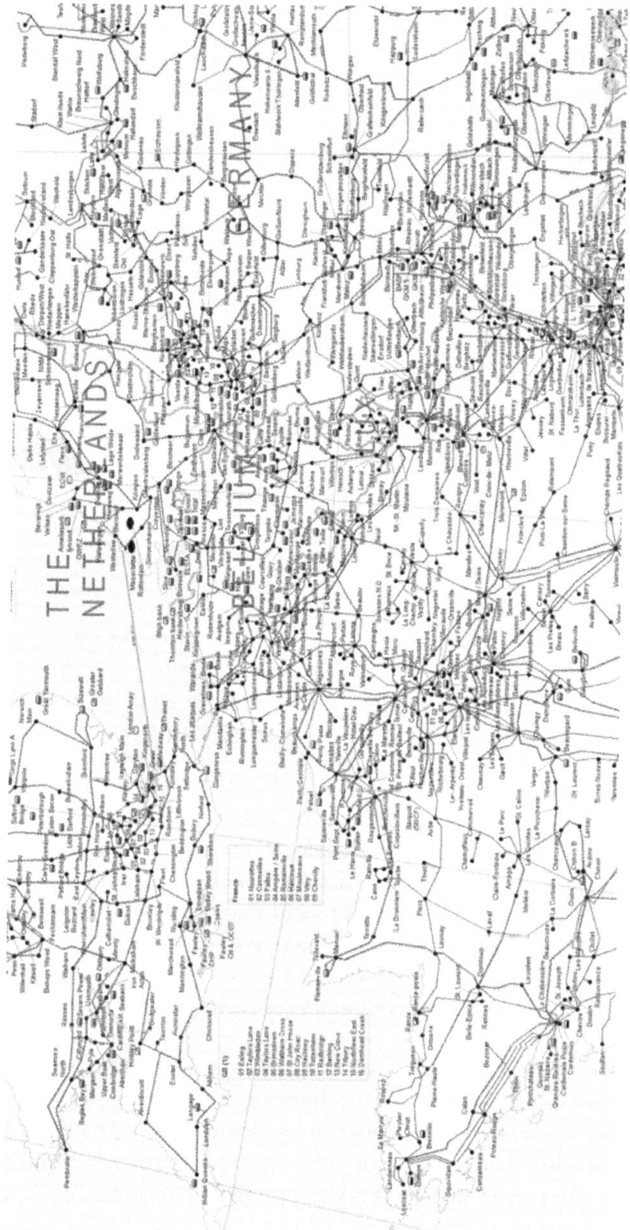

Figure 7.5. *Interconnections of European systems*

The underlying question in this type of evolution, which was generally very rapid given the intrinsic complexity of the elementary equipment, is how we organize these thousands of pieces of equipment into a coherent whole that is capable of surviving the dangers of natural disasters such as the violent storm in 2000 when a large portion of the power supply infrastructure was damaged and many households were kept "in the dark" – without there being a general failure of the overall French electrical grid, however.

REMARK 7.2.– At the scale of the entire grid, one nuclear power plant is considered an elementary piece of equipment. But taken as such, it is also a system that adapts to the needs of the network. This shows the fact that the notion of the system is a relative one that depends on the level of abstraction for the perspective we adopt. For the person managing the grid, a power plant either produces energy according to a known and predictable production rate or it does not.

For this to be possible, the addition of a piece of equipment in the process of growth must not be accompanied by a reduction in general reliability (*GR*). Without this, and in light of a well-known reliability distribution:

$$GR = REq_1 \times REq_2 \times \ldots \times REq_N \xrightarrow{\text{large } N} 0$$

since each REq is strictly < 1.

On the contrary, the addition of a piece of equipment must improve the resilience of the overall structure. By using a notation for "fixed points", we should be able to write: $GR(ES \otimes Eq) \geq GR(ES)$ in which *ES* designates the overall electrical system and *Eq* designates the equipment that will be integrated. We can imagine that this property is not easy to validate at the scale of the entire European network. The system's composition logic based on its elementary equipment, which controls reliability, is part of the overall abstract model that we call a "grammar" here, and for which we once more see the central role it plays[7].

We can formulate another observation that is typical of systemics.

IMPORTANT.– To survive, evolve and ensure its service contracts are met, any system must have a model for: (a) its internal organization, (b) its constitutive elements, and (c) the relationships between its elements so that any action taken by the company to ensure the continuation of its service contract over the course of its life cycle respects the invariable aspects that ensure its coherence.

7 See section 3.1 in Printz (2019).

In a system of this size, the elements must be organized, and therefore so must the models, such that the interactions can be controlled. If "everyone talks to everyone", the engineering will have to manage a complexity that will grow like the square of the number of elements, or more exactly their complete graph, that is $\frac{N(N-1)}{2}$ connections working in both directions, or $N(N-1)$ unidirectional connections. The solution is well known for all organizations and complex systems: the organization must be hierarchical, depending on the constants of time and reaction delays[8]. In this case, the complexity grows, as does the depth of the hierarchy, that is, by $O(\log_{Base}(N))$. In the case of the European network (Figure 7.5), we must distinguish national grids and/or those that come from different regulation regimes; there is only one regulator in France, the *Réseau de transport d'électricité* (RTE [Network for Electricity Transportation]), but Germany has several.

In the case of the French ES, there is a hierarchy of three levels (Figure 7.8):

– the point of usage, ideally entirely automatic, for distribution (some are still manual);

– the regional systems (SRC), of which there are seven (URSE, Unités régionales du système électrique [Regional Units of the Electrical System]), providing the first level of compensation within the managed region;

– the single national system (SNC), compensating among the various regions and providing exchanges with other countries via the CNES, the *Centre national d'exploitation du système électrique* [National Center for Use of the Electrical System].

All of these systems work 24 hours a day, seven days a week, alongside emergency systems in the case of a blackout to make sure the service is continuous, regardless of the cost of blackouts, since a certain level of risk must be accepted in all cases. The SRC and SNC are semi automatic systems in which the human operator can take over when there are conflicting demands and/or unforeseeable situations.

With the ES, we are confronted with one of the fundamental properties of systems as Norbert Wiener had imagined them: the double trajectography/dynamic (see Figure 7.9).

On the contrary, we have the demand from users which depends solely on their needs at any given instant *t*; this need is eminently variable, in quantity and/or in quality, from one moment to the next or from one day to the next, but it can be

8 See Chapter 8, "Architecture of complexity", in the book by H. Simon, *Sciences of the Artificial*, already cited.

observed and measured by ad hoc counting devices (the role assigned to radars in shooting systems). On the contrary, we have a function of offer that itself only depends on the state of the ES and its technical infrastructure at instant t. This state is well known but subject to unforeseeable hazards that can also be observed and measured. We should note that a system composed of N interconnected pieces of equipment, each with k possible states (working, stopped, standby, failure, etc.), itself has a number of states that grows exponentially by k^N, and is therefore time-consuming to calculate and difficult to validate.

REMARK 7.3.– At the European level, an analysis of network complexity means that to calculate the consequences of a line coming down (there are approximately 15,000), we would have to analyze and calculate something in the order of 300 million variables. The scale itself is a fundamental problem.

In both cases, we can make predictions in order to anticipate the probable evolution of these two functions and act so that they are as aligned as possible: that is the purpose of the ES. In a nominal (steady state) regime, this alignment is perfect since it is subject to the laws of physics.

The difference with the airplane in Wiener's shooting system at MIT is that in the case of the ES, these two functions are much more complex. The spaces where they are used are abstract spaces with many dimensions; but thanks to the available mathematical tools, we know how to model them. The transfer function requires a long description (see the notion of algorithmic information in complexity theories in Chapter 4 of this book), but we are able to get a computerized representation that is more or less complete, thanks to our knowledge of the laws of electromagnetism and thanks to the collected data which provides us with relevant statistics regarding the behavior of users, as well as climate hazards, failures, etc. The overall state is calculated every half hour in the best of cases. As with the shooting systems, the human element remains indispensable for fine-tuning, but overall it makes for very few people at the wheel for a system of such great size, managing a colossal amount of energy (100–110 gigawatts at peak hours); this shows that when we have the knowledge, if we want to do something, we can! Without the knowledge, we cannot do anything, but this, for engineers and scientists, is nothing new.

Behind this issue for engineers hides one of the most difficult mathematical problems that the ES allows us to identify. It is the matter of making two bundles in a particular trajectory, the ones with the offer (the 75 nuclear reactors, thermal power plants and hydraulic ones) and the ones with the demand from the 30 million clients (individuals, companies and professionals) coincide, in the topological sense of the term, even though they follow different types of logic, as we have seen. The small quantity ε of the vicinity is defined by the allowed tolerance for the balance of offer/demand. If we reason in a 3D spatiotemporal space, we not only have to connect the points of the two curves of the bundles at instant *t*, but also the tangents and the accelerations, which in a 3D space requires a focus on the curve and the torsion of the trajectories, or three vectors and nine coordinates attached to each point. Since each element of the system has its own life, we understand that the general balance, in which the one compensates for the other, with time steps that are less than a second (see Figure 7.7), is not an easy scale to regulate. We also understand why companies like EDF and the RTE must maintain and develop skills in modeling and simulation at the best mathematical level possible, knowledge that depends on simulation tools which must be mastered and entirely understood if one is to avoid getting misled by models; we will come back to this in the conclusion when discussing future problems. The computer models that are the result of this, properly validated with respect to physical reality, enable us to guarantee the proper functioning and stability of the system (it is a dynamic balance); they are used on a daily basis by users of the SNC (at the national level) and the SRC (at the regional level) systems.

7.2. Interoperability and cooperation

To grow and develop in the living world, cooperation between organisms seems to be the rule. To exploit the energy resources of an ecosystem, nature imagined two families of solutions: either many small organisms everywhere on the territory, or very large organisms like elephants or whales, or previously, the giant reptiles of "Jurassic Park", spread over half-empty, vast territories. But it would appear that the solution preferred by evolution, one that is both more stable and more durable, is that of cooperation among small-/medium-sized organisms to form super-organisms like those which constitute a species, or organized collectivities as is found among social insects. A species carries on well beyond the life of its individual elements, and in this domain, it would appear that insects have the advantage in terms of longevity[9]. Many small, cooperating organisms are clearly a better guarantee of survival. A very large size seems to be an obstacle that evolution has eliminated. Biomechanics and bioenergy have their laws, which, associated with excessive specialization, make adaptation problematic for large organisms when there are large-scale changes.

All things being equal, we find analogous phenomena elsewhere, with the huge machines we are capable of building, at least in theory. A "huge" machine means a large quantity of energy is produced/consumed that we must then know how to regulate. We have known since almost the time of Carnot/Clausius that thermal

9 For an in depth analysis, see the testament by S.J. Gould, *The Structure of Evolutionary History* (Gallimard, 2006), in particular Chapter 8, "Species as Individuals in the Hierarchical Theory of Selection".

devices have an output that improves with the quantity of energy produced and the size of the machine (in this case, the production of high temperature/pressure vapor), hence the impressive dimensions of turbine/alternator groups in a nuclear power plant; in this very precise case, "small is not beautiful". But an unbalanced, large machine (a formidable problem in engineering) will produce vibrations (one of many ways of dissipating lost energy) that can amplify through resonance phenomena and become concentrated to the point of destroying some element of the machine. In the case of the latest generation, the 1.3 gigawatt turboalternator group, this ends up in breakage among the mechanical pieces that can destroy the machine! In production, there is thus a maximum size that cannot be surpassed, and this depends on the available technology, the quality of the materials and the state-of-the-art expertise in engineering.

REMARK 7.4.– With the latest generation natural gas facilities, this is less the case, since the power can be modulated over a rather large range without altering the price of KWh.

One of the immense advantages of electrical engineering for users is that an entire range of machines can be built, from very small to very big, and therefore, energy consumption can be dosed in terms of the energy needs of individuals, and with excellent yields. There is thus a structural dissymmetry between production and consumption given the laws of physics and behaviors.

In the natural world, the most common cooperation strategies are based on either (a) the sharing of one genome within a species, or (b) the development of symbiotic relationships, that are mutually beneficial, between different species. For human societies, it would seem that the cultural element plays a very large role, as has been shown by certain sociologists and philosophers. As a humorist once said, "culture is what remains when everything else is forgotten", but the line is much deeper than it appears. Culture has the same function as a behavioral model. It is the "language" of the common cultural property that is acquired over a long process of education before it can act efficiently in a collective way. In this sense, it is an element of a "social grammar" that organizes society[10]. A language that is poorly assimilated is a collective risk caused by an individual who was badly educated and/or who deliberately violates the rules. The term "corporate culture" is part of the vocabulary of proper governance, and in the case of companies in charge of collective goods such as energy, railways, communications, etc., the term is not an idle word.

On the user end, we are instead in a type of symbiotic relationship, a co-evolution that amounts to the purchase of a service, in exchange for the provision

10 See, for example, Bicchieri, C. (2006). *The Grammar of Society – The Nature and Dynamics of Social Norms*. Cambridge University Press.

of energy according to a service contract negotiated between the consumer and the provider. We obviously cannot ask the user to understand the model for the system they are using, but they must nevertheless respect certain rules stipulated in the service contract. Likewise, we do not ask a motorist to know the mechanics and the computing involved in using their vehicle, but here also, they must respect rules such as the Highway Code and maintenance of the vehicle.

All of these considerations are found in artificial systems, and in the electrical system in particular. Indeed, to serve a maximum of users in an equitable way, taking the problems of machine size and organization into account, the most natural solution is to share the technical aspects of the ES and engineering teams. In other words, the electrical AU that we previously spoke of must be made interoperable. We thus reach a situation where there is only one single community of users, but with different needs and a transport network that carries the offer of all the pooled means of production on demand. The transformation that occurs is summarized in Figure 7.6.

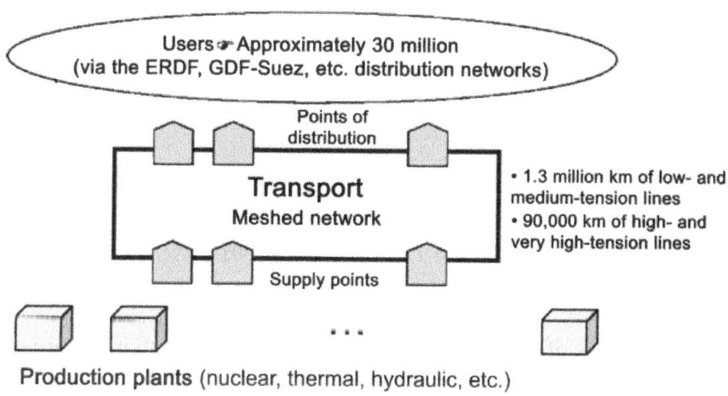

Figure 7.6. *Interoperability among means of production and transport*

In this approach, a user does not know where their energy comes from. The transport network ensures that a service is offered to users in general by optimizing the mechanisms for compensation between the available means of production, as discussed earlier. This works according to the laws of Kirchhoff. The network mesh makes it possible to define several transport paths from the point of production, in general, towards users, and therefore makes it possible to improve the service contract; and this is now also possible at the European level. Users are connected to the transport network via a distribution network that was initially managed by EDF, although today approximately 155 operators exist, the principal ones being ERDF, GDF-Suez, etc. If we reason in terms of models, we see that one of the major problems of the transformation is creating coherence among models for each of the initially independent electrical AUs and their interactions. The transformation of the ES organization has gone from an architecture in silos to a hierarchical architecture in layers.

To enact this type of solution, interoperability amounts to coherently taking everything from families of specific constraints to different elements of the technical system into account, all materialized in different models:

– Organic constraints tied to the nature and physics of the constitutive pieces of equipment in the ES, for example, a High Voltage line cannot carry more than x KW, otherwise it will melt!

– Project engineering constraints, since what was a (relatively) “small” project in the case of silo architecture will become a “large” and complex project in layered architecture, this time accounting for the limits tied to the size of projects that we are capable of completing; 10 independent projects of 100 people each is something very different from a single project of 1,000 people. We must therefore demonstrate “agility”, which is the price to pay for the change in architecture.

– Usage constraints which mean that users are required to respect rules that they did not need to worry about when the electric AUs were independent. The system can protect itself from maliciousness, but only to a certain extent.

To cite just one example, we can mention, in terms of physical systems, the establishment of standards for the voltage and frequency of the current for users, 380/240 three-phase voltage in 50 Hz, with an interprofessional agreement for all actors of the electrical system.

The common rules of these new communities will affect many more people; they will necessarily grow somewhat in number and it will be important to monitor them so they do not become contradictory, which is a difficult logical problem when they become numerous. The principle of simplicity, the famous Ockham’s razor, must be respected diligently so we do not lose control and so, in the end, the ES and its users do not transform into an unmanageable Babel. The rules of survival for an entire species are different from the ones that apply to the individual members of the species.

The logic of cooperation is of another nature compared to the one that determines the behavior of a lone individual. It is founded on trust. We could also cite the Bellman theorem, which says that the overall optimal state of a system is not the sum of the localized optimal states for each of its elements. The logic of cooperation requires actors who are more virtuous, whether they be the users or the engineering teams; this is the price to pay for the comfort that establishing common resources brings. In the case of engineering, it is a matter of the fragility of the approach since learning and maturation (the role of quality control) are medium-

/long-term phenomena that can be incompatible with a financial logic that always favors the short term.

In the case of the ES, equipment such as the EPR reactors being constructed in Europe and in China are designed to have a life cycle of approximately 100 years, which means that the people who designed them will no longer be there to see them dismantled at the end of their lifetimes.

In conclusion, we can see the emergence of another observation that is typical of systems approaches.

IMPORTANT.– Seeking interoperability translates, concretely, into creating shared models. This means that the models must be designated as such, not implicitly in the head of some actors who are consulted as if they were oracles. The common, abstract model of these different models (in computing jargon, we say the "metamodel") must be known by all interested parties, and which aspects are relevant to each, so that all actors can act according to the logic of the model organizing exchanges.

This abstract model functions as a language spoken by the actors of the system: it defines the grammar of the system, that is, its reference framework. It is the obligation of the actors, especially the methodological "grammarians" in charge of its definition, to consider it a common good that requires all their attention since its coherence is fundamental.

The model must be distinguished from documentation written by just anyone, which incidentally must respect its rules.

7.3. Resilience

We will take the word resilience in the literal sense, which is a measure of the capacity of a mechanical part to resist shocks, that is, its ability to absorb the energy of a shock while continuing to provide the expected service without breaking. The transposition of the concept into the world of systems that are subject to the hazards of the environment is natural.

We have just seen, in the previous section, all of the advantages tied to the pooling and sharing of resources to offer users a better service. But sharing the resources also means sharing the hazards. Consequently, what was previously a local incident or breakdown, affecting one access point or one region, can now affect the entire ES, including all of Europe, given the interconnections (see Figures 7.5

and 7.8). This must be avoided at all costs, for if this is not the case, then sharing will be a failure.

One of the great successes of ES is that not only has pooling been successful, but the service contract has also been improved thanks to the compensation mechanisms whose development was made possible by the meshed network. If one resource disappears for any reason, but another equivalent resource exists elsewhere and is currently available, then the fault can be compensated for. In fact, the overall network functions as a kind of insurance because it was designed that way. We share the risks incurred by users because the network, overall, is not subjected to hazards all at the same time; it is, in fact, the network itself that absorbs the risk and manages it.

The meshed transport network is the seat of intense energetic phenomena since it can potentially carry all of the available production, approximately 100 gigawatts, and even a little more when it is very cold and supplies are needed outdoors. To give a point of comparison, a 35-ton truck driving at 110 km/h releases, in the case of a collision, an energy equivalent to 16 MW, or 80 trucks per latest-generation power plant (1.3 GW), and more than 6,000 for the entire production. It is the equivalent of a giant machine that needs to be carefully regulated to manifest the fundamental equilibria such as the ratio of offer/demand. For example, if an artery of the network no longer works, for whatever reason, the corresponding energy is almost instantly transferred via other equipment that must therefore absorb the equivalent of a shock wave.

Figure 7.7 provides an overview of the hazards over time, and against which transport network managers must stand guard. For example, a lightning strike is a very quick energy phenomenon, less than a thousandth of a second, that can destroy equipment, thus putting it out of service. A sudden blackout can cause destructive power surges and even form electric arcs with very high tension. In northern countries such as Canada, certain electrical lines can be affected by magnetic storms generated by solar activity due to the proximity of the North Pole.

Nevertheless, certain phenomena can be anticipated: a storm does not start when the weather is clear, there are always forewarning signs, and the weather forecast gives projections and storm alerts; it is the same for solar activity (electromagnetic waves take approximately 15 minutes to travel the distance from the Sun to the Earth, but the particles that perturb the magnetic field on Earth are much slower). The role of a forecast is essential for managing hazards, along with the measurement instruments that are indispensable for observing what is really happening. For example, the temperature of equipment is a precious piece of information, as is the level of water in retention dams or even the weather forecast for local temperatures, etc. The management of corresponding information and the maintenance of serious statistics let network managers act, fully informed and make simulations to understand what is happening in real time (something that can require a great calculating capacity), or what has happened, thanks to delayed simulations, not in real-time.

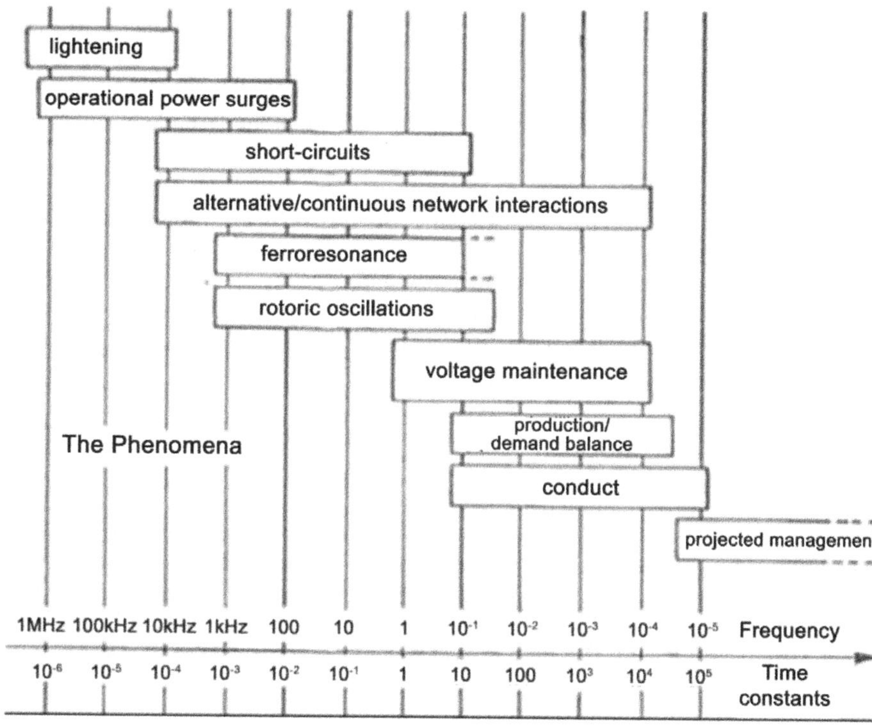

Figure 7.7. *Time constants of the transport network*

Once the breakdown has undergone diagnostics, a security perimeter must be established in order to prevent the shock from propagating through the rest of the network, and the information must be sent to the network managers. This is a well-known, classic mechanism, which can, however, be difficult to put in place given the speed of certain phenomena. At that point, all that is left is to make the most of the inertia specific to the network equipment. For example, a High Voltage line that absorbs a power surge caused by a rupture in another line will begin by heating up, the cable will expand and approach the ground until it sparks a destructive electric arc or melts, etc.; but all of this will take a certain amount of time, during which the managers will be able to act and organize selective power cuts, accounting for service contracts for different clients. The precise knowledge of the behavior of equipment in extreme situations, outside of normal functioning, is indispensable if the managers of the network are to properly trigger their responses.

This knowledge is materialized in the form of one or many models that constitute the prerequisite to any serious computerization; they define the set of possible actions, that is, the "grammar" of the system, in the linguistic sense of the term, a structuring aspect nuanced in this book. It is a domain where simulation can provide precious information for decisions that need to be made within the space of a few minutes, depending on the situations encountered.

This feedback from experience is indispensable for properly understanding new problems that regularly affect the network in its entirety.

The map in Figure 7.8 shows the organization of regional frameworks defining security perimeters. In the case of a problem, the regions can be separated from each other. This map shows the management size and capacity limitations that must be taken into account if the ES is to remain generally manageable and a maximum number of homes with power. This information is central to understanding the logic of modeling and the associated models.

There is an important lesson to learn from these mechanisms in terms of systems organization in general.

IMPORTANT.– The nature of the phenomena that a system can be confronted with must be understood as much as possible, and the cognitive biases that are always possible among human actors must also be considered[11]. Ensuring the proper functioning of the ES has the same role as an immune system, all other things being equal: to protect the system against environmental hazards and internal errors.

This holds for the environment beyond the system, the *outside*, an environment which can always be observed. In a case where knowledge about this environment is lacking, it is the responsibility of the observer, since without observations, and without real measurements, feedback from experience becomes impossible, or else it must be the work of magic. The behavior of users must be able to be analyzed. They have rights as well as duties.

This also holds for the system's interior, the *inside*, an interior that always includes two components: (1) an organic component, with material equipment whose physics and behavior are known, by definition, although knowing how to use the corresponding information is another matter, and (2) an organizational and human component that deals with engineering and whose limits are known. The first task of the management is to ensure that the personnel in charge of this engineering have the required expertise, and that the organization is properly playing its part, for

11 This is a full-sized illustration of the fundamental map/territory relationship recalled in Chapter 1.

example, by not hiding problems. This is the price of having the trust of system users.

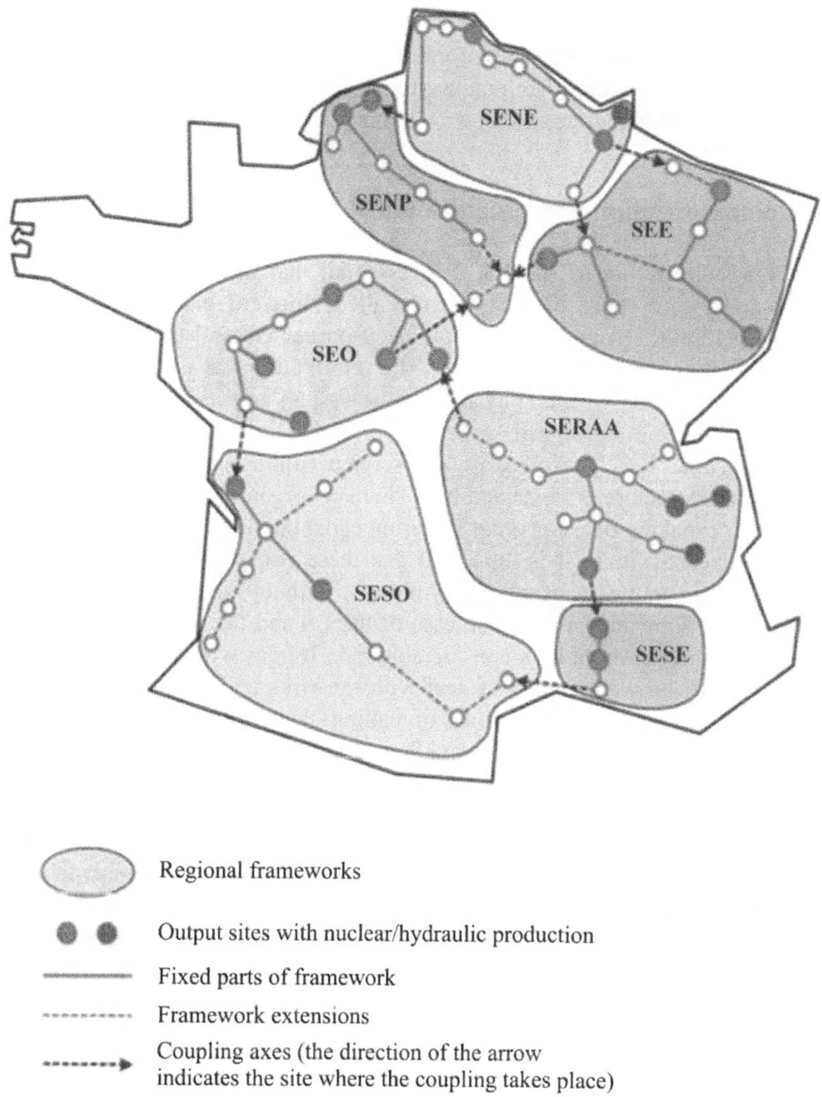

Figure 7.8. *Regional frameworks and security perimeters*

Another lesson is that of quality – in the sense of relevance – of collected information. This quality is essential to understanding phenomena, whatever they may be, including organizational and human phenomena like the ones that take place in the context of experiential feedback (blackout rate), or the analysis of user behavior, including malicious acts.

IMPORTANT.– Not having relevant information, which is something that threatens the entire quality chain, makes the organization of this type of system impossible, hence the role of information systems that we will now discuss.

7.4. Computerization and organization

In a system as big as the ES, it is essential that information concerning the current and future states of the system (called "forward planning" among those working at EDF/RTE) goes back to the decision centers and the system managers (production, transport and distribution). The information that is available/accessible to managers depends on the "physics" of the system and its organic structure, the latter of which is optimized with the primary criterion of ensuring proper functioning and, in a general way, the service contract (i.e. the task of quality control). This structure is replicated by a "human" organization that must work like a "machine", making decisions (not flattering egos) that are the product of its sphere of control/specialization. It is thus vital that these two structures are "aligned" so that they act in mutually beneficial symbiosis, without which decisions would be inappropriate, something that in the case of the ES and the management of intense energy phenomena would be a considerable risk. It goes without saying that all the managers must be well trained to deal with the risks and their decisions must be uncontested, without the possibility of calling them into question. The quality control culture in this context is not an idle word.

As in the shooting system that was the object of Wiener's first studies, we must distinguish between reflexes that are beyond human abilities, the reflex time belonging to the system, and those which come from human decisions, the time of reflection; with, in between, intermediate reflexes that depend on human performance (e.g. piloting a combat airplane or a Formula 1 car requires exceptional physical abilities; commanding a military operation requires understanding the capabilities of men and materials as precisely as possible, with great clarity, a resistance to stress, including the stress generated by the politics of it all, and real courage).

A turbine/alternator group of 1,300 megawatts turning at 3,000 RPM to stabilize a frequency of 50 Hz requires reflexes whose reactive loop is in the order of the millisecond, completely inaccessible to any human. This is the same for the case of anti-thunder equipment (see Figure 7.7). On the other end of the timeline, provisional guidance and management require a time of reflection since it is necessary to evaluate the impact of decisions depending on the real situation at instant *t*, since several choices are generally possible.

The capacity of Science and Information Technologies has obviously played a major role in the evolution of systems and their organization. Let us provide two examples.

The reliability of data networks makes it possible to transform online control into remote control, which in turn makes it possible to better manage a larger set of resources up to an optimal size that integrates the limits of human capacity. This makes it possible to separate control in the concrete sense of onsite operations, which improves decision-making.

The processing capacity of computers makes it possible to (a) provide much more detailed provisional management, and even real-time simulations in certain favorable cases, and therefore a better use of energy resources, whether thermal, fuel and gas or hydraulic, a resource that must be conserved since it can be mobilized in a few seconds, and (b) better distribute the load on the network.

The software component of ES therefore has an increasingly important role, with the so-called "intelligent" transport network, or the *smart grid*, in mind which is indispensable for correctly managing the integration of renewable energy, which is now widespread and random in nature, into the operations of a system based, let us remember, on the balance of supply and demand (among others), something which everyone can understand. The principle is summarized in Figure 7.9.

The dichotomy of *inside/outside* is materialized by the two curves, Evolution of Demand and Evolution of Supply, curves which should ideally be perfectly superimposed even though the hazards they are subject to are very different; let us recall, as we said previously, that we are dealing in families of curves. The fluctuations in demand become more random: it depends on the decisions of each of the 30 million users. Fluctuations in the supply are more structured, depending on the workloads of each production unit (PU) but subject to environmental hazards notwithstanding weather models and their uncertainty, equipment failures and for some years now, terrorist threats.

The size of the system and the basic regulation cycle mean that the volumes of data circulating in the communication lines can be colossal. For example, if we wish to take samples every minute of the demand among 30 million users via "smart meters", the IS will have to process approximately 43 billion messages per day. It is therefore absolutely necessary to subdivide them so that the user data can be combined for comparable profiles. In doing this, the system thus reasons according to an "average user" who can reflect the specific needs of individuals.

As for heavy equipment, there is much less of it than there are users, however the situation is not better since much more complex states need to be tested with much more detail, and the response time must be guaranteed. A piece of equipment which is overheating must offer retroaction within a timescale of about one minute in order to remain functional, something which can quickly become incompatible with processing capacity and/or data transmission speeds. Certain blackout decisions can require human intervention, at best within a couple of minutes, time which the system must take into account.

What the ES illustrates perfectly is the fact that centralized “intelligence” is not sufficient to ensure the service contract is honored. Its centralization is in itself a risk since all systems depend on the “center”. Decision-making should be done by subsidiary and intelligence should be “distributed”, but not in any manner, especially not randomly, as indicated in Figure 7.8. The right solution is the cooperation of regulation processes via information exchange so synchronization can happen when needed; this is the only way to “break” the combinatorial and organize it in a way that is manageable to humans.

The ES on its own shows the inanity of the absolutely deterministic postulate formulated by Laplace in his *Traité des probabilités*:

> We should thus imagine the current state of the universe as the effect of its previous state, and as the cause of what will come next. An intelligence that for a given moment would know all the power that drives nature and the respective situation of the beings that make it up, if it was however vast enough to subject this data to analysis, it would envelop in the same formula both the movements of the most grandiose bodies in the universe and those of the lightest atom: nothing would be uncertain for it, and the future, like the past, would be present before its eyes.

And yet Laplace had such a great mind! Without realizing it, he was doing metaphysics[12]. His postulate was simply false, or more precisely, empty of meaning and inapplicable, unless we suppose that the information is sent and processed with infinite speed and costs nothing. So, be skeptical and be careful when you are confronted with real, complex systems issues, especially when they touch on the outside world, where everything has an energetic cost, including the information needed for regulation. It is the paradox of Maxwell’s demon, unknown at the time of

12 Naturally, we cannot fault him for not knowing what would be discovered a century later by Einstein!

Laplace, who plays such an important role in matters of information. Compared to the energetic scale of the universe and astronomical phenomena, the ES is but a small object.

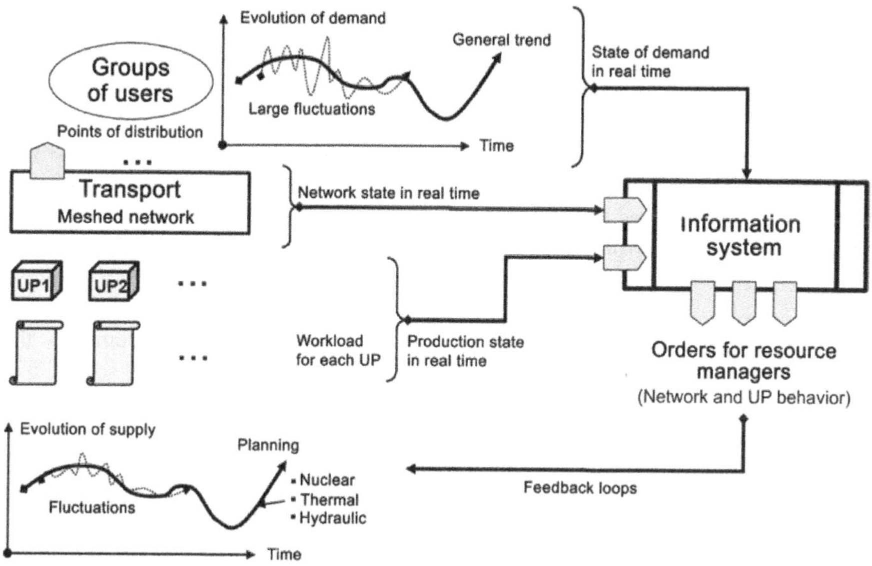

Figure 7.9. *Principles of computerization*

Another lesson from ES is that to ensure the service contract is respected, intelligence must be distributed as close as possible to the decision, so that the feedback loop can be compatible, on the one hand, with local constraints, and on the other hand, with the general constraints of the ES. A decision that would save the service locally by keeping the rest of the system "in the dark" would definitely be a bad decision. But then, how do we distribute intelligence in a coherent fashion? Where do we place the cursor?

First, by creating a hierarchy, as we saw in section 7.1 (national level, regional level and points of access), but especially by deliberately distributing the general rules to follow all of the elements of the hierarchy, which are necessarily heterogeneous. This is another illustration of the strength of models that reflect the overall logic that all the active units in the ES must follow. Ideally, any node should be able to play the role of the center; they must therefore all have access to the overall abstract model, even though locally they only use one part of it.

If we have truly understood von Neumann's message, "who keeps watch over the watchmen?", we have to make sure models are used properly, at the proper moments, and establish a control system that makes sure of this, which – in order to escape the paradox of Laplace's centralizing demon – must necessarily be distributed, as a kind of demon, but integrated into each node of the system, and whose only task would be to constantly ask, "am I doing what I was tasked with doing correctly?", perhaps by going to see what a neighboring node does! If a rule is broken, for whatever reason, the alarms should be sounded in the right places… and drastic measures of isolation and reconfiguration should be taken. We are thus very close to the biological phenomenon of apoptosis, described by biologists[13]. The active unit that violates the common rules is not part of, or no longer part of, the system; its probable error condemns it and it must be isolated, confined and then cut away from the system since now the survival of the whole is at stake.

The most surprising thing about the computerization of systems is that progressively, this computerized part of the system comes to represent a sort of dual, "informational", virtual image of this system, an image of which a part is precisely an instance of the model of the managed equipment and energy exchanges. This new abstraction – which is not easy to detect at first sight since it is an immaterial, logical "form", a set of relations – emerges progressively from its material shackles, thanks, in particular, to the interoperability constraints specific to Information Systems.

In Figure 7.10, we provide a real image (created at the start of the 2000s) of this fundamental evolution (a bifurcation, a point of no return), at the moment when it was realized that the evolution of the device needed to be revisited, a device where the need for information exchanges translated into computing links between the interested parties[14]. This cartography represents a volume of code of about 20 million source lines, or approximately 10 million hours of work on its development (excluding maintenance) in "gross" units of work, but three or four times more than that in "adjusted" units, given the surprises and rate of success in projects according to the statistics of the Standish Group[15]. Very quickly, the proliferation of the number of links that merely translate the need for exchanges became a hindrance to the evolution of the system as it was a combinatorial of, at best, N^2 … and some time for reflection needed to be taken.

13 See https://en.wikipedia.org/wiki/Apoptosis for more information.

14 By an RTE team, in connection with teams from the DER in Clamart, for whom I was lucky enough to work as a consultant. This was the R&D organization of EDF/RTE.

15 See the website: http://versionone.com/assets/img/files/ChaosManifesto2013.pdf.

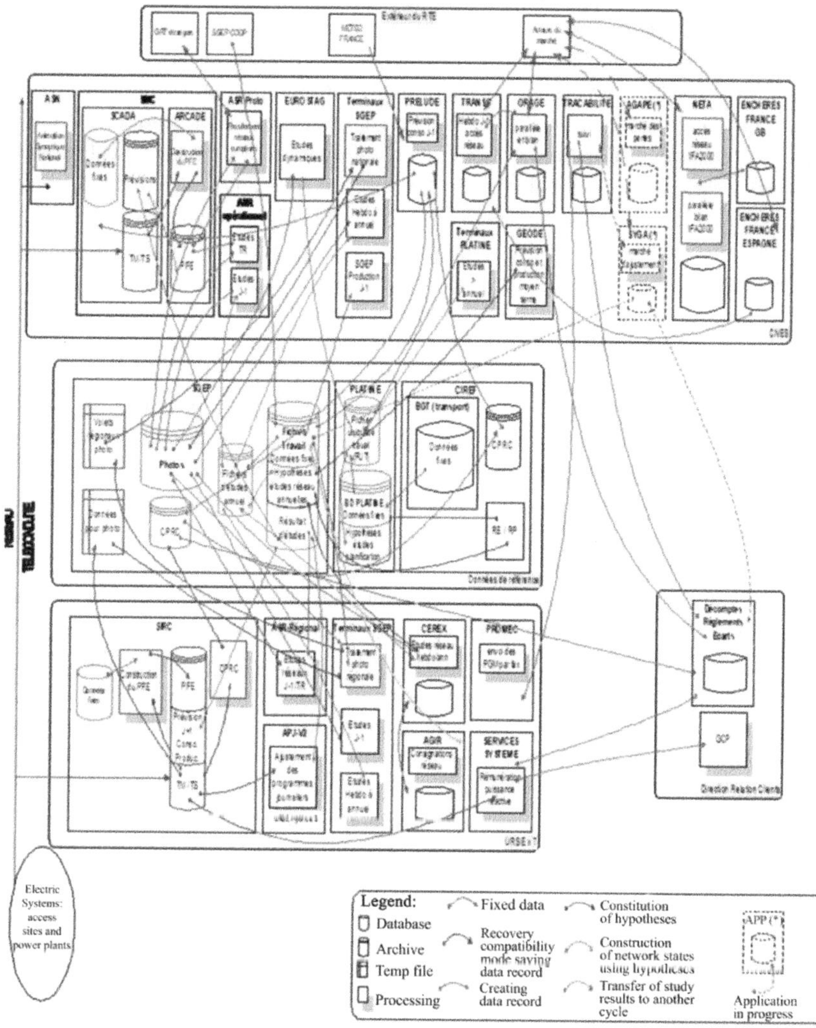

Figure 7.10. *Progressive complexification of the IS of the electrical system*

The physical transport network doubles as an information transport network in the form of generally asynchronous messages that must respect deadlines, which in the jargon of programming is called a databus and is the means by which exchanges are exchanged, in theory, including all the information exchanges between equipment. The architecture of this bus is the exact image of the exchanges that were previously physical between the elements of the system, depending on their organization.

In this respect, the organization of the bus is a fundamental element of the model for understanding the overall workings of the system, something that corroborates the idea that a programming language is in essence a language for specifying models, a language that can be "reworked" as such, thanks to computing technologies (such as the theory of languages, abstract machines, etc.), but never losing sight of the end goal of this vast ensemble. For this, a "semantics" of actions to be taken must be defined using an abstract machine that can be created on any commercial computer (operational semantics). The software bus plays the role of mediator by interrupting the complexity that end-to-end coupling had created over the course of successive adaptations (from the point of view of electricity, they are short circuits).

The most remarkable point is that the computer image obtained in this way is the exact reflection of what we must know (both necessary and sufficient) to interact correctly with the system. It is a grammar that makes it possible to speak the language of the system while respecting the constraint of inside/outside, an old idea among the logicians of the Vienna Circle which von Neumann made his own by putting it at the center of his architecture and his logical design: we must distinguish between the internal language of the computer (in this case, binary code) and the external language made for humans, with the understanding that both must be consistent and complete[16]. John von Neumann had started this theory in his studies on reliable automata. In doing this, we break away from *inside/outside* evolutions through a purely logical device, immaterial, all the while remaining coherent, which is an essential property for the maintenance and scalability of the system. Thanks to computerization, it has become possible to go from a logic of links between points whose complexity is $O(N^2)$ into a logic of linear complexity, notwithstanding the creation of a pivotal structure that concentrates the complexity in one precise place that can be organized in an ad hoc manner, rather than letting it spread and metastasize everywhere, which would make it unmanageable. This is what the teams at the RTE did when they added integration middleware.

This device is an illustration of the notion of the transducer, whose importance was recalled in Chapter 2.

7.5. Future problems

From a strictly systemic point of view, without entertaining considerations of energy policy and politics that would be extraneous at this stage, we can nevertheless now point to some problems that will need to be resolved if we wish to sustainably maintain the capacity of ES at current levels.

---

16 See, in particular, Carnap, R., *The Logical Syntax of Language*, Routledge & Kegan Paul, reprinted in 1971.

7.5.1. *Renewable energy and the optimization of the network at the European level*

The introduction of renewable energy such as solar, wind, etc. into the ES makes supply planning more difficult since these types of energy depend on the vagaries of localized and particularly unstable weather[17]. The fluctuations indicated in the drawing in Figure 7.9 are amplified. The induced effect of renewable energy is that initially only the demand was unstable, whereas now, the supply is unstable as well. To preserve the balance of supply/demand in satisfying economic conditions, testing to understand the nature of the supply must be done with much more precision, which has a direct impact on the information system: quantities of data and processing capacity in much more real-time. Planning the workloads (see Figures 7.7 and 7.9) according to provisional management every evening for the following day (on a J – 1 basis, as they say at the RTE, where J stands for *jour* or day) is inadequate for this type of energy. In terms of regulation, immediately available energy resources must be reserved, and they can only be thermal or hydraulic. With hydraulic energy being limited by geographical and environmental constraints, only thermal energy is left. The choice could therefore not be any simpler: power cuts must be increased or the thermal infrastructures must be expanded (and much more even, if nuclear energy is abolished).

Incidentally, we no longer control the distance between the location of production and the location of consumption. Now, solar and wind maps must be taken into account. The sun is more likely to be found in the south, and perhaps even in deserts, in the places where people cannot live. Wind is more likely to be found in the north, even the far north, where the winds are regular and powerful. But populations tend to be concentrated in the temperate zone, in the places where it is nice to live. Carrying energy over long distances, with current technologies, augments dissipation and losses on the line because of Ohm's law. The problem is therefore not at all easy to resolve, requiring long simulation studies to better understand in which direction we will proceed in order to master this transformation. To speculate on the ignorance of users to promote economically unreasonable solutions is to create a potential social risk that could implode at any moment.

Renewable energy, coupled with issues of deregulation in the European context, make problems of network optimization and the concomitant complexity of this optimization clear.

A first remark, due to the fact that production is no longer necessarily done as close as possible to the zone of consumption: the *inside/outside* boundary, whose

17 On this current problem, see the works by Jean-Marc Jancovici: https://en.wikipedia.org/wiki/Jean-Marc_Jancovici.

importance was previously underscored, becomes porous. Consequently, this greater quantity of energy circulating in the network will need to be managed more attentively to avoid the risks of a power surge. A wind and/or solar farm produces energy, whatever the demand, leading to the misguided, counterintuitive idea that it is better to use it to "save" the non-renewable energy, whether thermal or otherwise. Whereas wind is relatively predictable, with a margin of about 10% according to current statistics, solar is, on the contrary, largely unpredictable, all other things being equal. A slightly cloudy day can make the yield of an installation, which in any case follows the cycle of the sun, fall drastically, but not the demand. On a European scale, currently, solar production is at about 25–30 gigawatts, which can fall to zero in the space of a few minutes. When we know that the available power installed in France is at most 110 gigawatts, we can measure the breadth of the problem caused by incertitude since, to avoid power cuts, power plants must be kept on reserve, just in case.

A second remark, due to the fact that in order to optimize – and this is not at all obvious for a non-specialist – we need information on the overall situation of the network at instant T. It is therefore necessary to exchange information, in great quantity and at high speed, between the different regulators. The energy produced, once connected to the network, propagates almost instantaneously towards the sites of consumption as long as the network supports the load. Therefore, if the quality of exchanged data is not guaranteed, this can induce erroneous decisions on the part of the regulator, which can be dangerous when we are close to the limits. We once again arrive at the question posed by von Neumann, "who keeps watch over the watchmen?", for in the case of a power cut, who in the end will be responsible, given that the event causing it could be very far from the zone left "in the dark"?

The "naive" desire to optimize the cycle of supply/demand by getting everyone on the same page, something which by all appearances seems like common sense, is automatically accompanied by a formidable increase in complexity, which will be materialized in the information systems of different regulators trying to organize this complexity. We must consider that what was possible in France between the regional SRC systems and the national SNC system thanks to the centralization and monopoly of EDF would not be transposable to the European scale where a European SRC is politically unimaginable given the current state of Europe, and would be very risky on a technical level given the state of our knowledge about the dynamics of very large systems with millions of variables of states.

We see, and that is at least at first glance just common sense, that optimization translates operationally into more fluctuations, more uncertainty and more unpredictable hazards. Therefore, to attenuate the risk of power cuts, an increase in regional capacities is needed, creating a layer of protection against external events, just in case. No politician who is worried about their popularity would have the courage to go and explain to their fellow citizens who have just experienced a power outage that it comes from a decision by European regulators to optimize the network. Having less costly energy, but with power cuts such is the dilemma.

The situation created in this case is precisely the opposite of the initially sought-out effect; it is, moreover, a magnificent illustration of Bellman's theorem on optimal control, which says that the sum of local optima does not constitute an overall optimum, and reciprocally so.

A third and last remark owes to the dissymmetry of the information needed for regulation in the general sense of the term. In the time before deregulation, EDF had "complete" information about the entire system, which was the advantage of the monopoly, and it had all the means of obtaining the information needed for the proper management of the ensemble. With deregulation, information becomes more and more imperfect, which requires the establishment of an independent regulation authority, here the CRE (*Commission de régulation de l'énergie* [Commission for Regulation of Energy]). A new problem is thus created, that of the expertise of those sitting on the commission: where do they train? Who will be the arbiter of conflicts, and with what authority? Are they truly independent? The problems are countless and the sources of conflict unsolvable, opening the path to a litigiousness that will only benefit lawyers, all of which will rest on the shoulders of users since it will create a new cost center.

At the level of nation states, we can imagine that compromises could be found, with a last recourse to a referee state. But what happens at the European scale? Information is, in the best of cases, imperfect, perhaps incomplete, but it can also be biased, or even manipulated as we saw in the ENRON affair. Who can guarantee that directors are saints and will never commit indiscretions, especially if their salaries are indexed on their company's stocks? In this case, even the notion of optimization itself no longer has any meaning. We enter a logic like that of the "invisible hand of the market", whose consequences we have recently experienced.

The size of the system, the uncertainties tied to human complexity, themselves tied to the number of organizational actors and their cultural diversity, the more and more extensive hazards related to the environment, the vagaries of interactions: all of this will create a horizon that, if we go beyond it, will translate into a loss of control. The controllable/uncontrollable boundary becomes unclear, and moreover, with the growth of complexity that results from couplings, it grows into a sort of no man's land where anything can happen.

IMPORTANT.– In conclusion, optimize, certainly, but to what extent? Only simulations can perhaps provide a response, given the state of our knowledge.

7.5.2. *Domestic photovoltaics*

Asking every home to install photovoltaic panels, wherever possible, is the same as asking everyone to be a potential producer. Either the energy produced is consumed or stored there, using batteries or any other compatible means for domestic use (compressed air, boilers, etc.), or it is delivered to the ERDF or another operator (there are more than 150); but then the network functions in the reverse direction. The local/regional distribution network needs to be reorganized.

The current produced in such conditions is of variable quality. If it is for heating a house or cooking food, it is not a problem. If it is for powering electronic equipment requiring a high-quality current however, inverters, which are costly, need to be added. We thus see a double power supply taking form: one for sensitive electronic equipment with an active service contract (telephones, computers, security systems, etc.) and another for everything having to do with domestic life (heating, lighting, etc.), similar to what we have with potable and non-potable water. Keeping in mind that a smart grid ipso facto means high-end electronic equipment everywhere, and also the need to maintain operational conditions alongside the service contract since, naturally, it will not work on its own all the time.

In any case, we can clearly understand that the overall problem is not as simple as it seems if we integrate all the collateral factors, including changes in the habits of 30 million users.

7.5.3. *Europeanizing the network: the problem of governance*

Europeanizing the network is a form of energetic "globalization" of greater and greater sharing at the European scale, which can intuitively appear beneficial for the user-consumer; but as we have already pointed out for the French case, this also means sharing the risks, and since energy shocks spread almost at the speed of light, this means that an incident in Denmark can cause a blackout in the south of Italy in the space of a few seconds. We thus need to harmonize the models for different ways of managing European networks, presumably under the authority of the European Commission, if it is competent in these matters, or via an agency in charge of European-wide regulation. This is clearly a gigantic, long-haul task that cannot be decided according to political criteria since the problem is essentially technical. The magistrate is the laws of physics, which cannot be rearranged according to moods nor repealed by parliaments after the fact. The work can therefore not be done by civil servants with hypothetical competence in systems and commissions of experts named by governments to supposedly "advise" them and whom we know to have a varying sense of ethics all too often. We now arrive once more at the analysis made by Jay Forrester who, after his experience with Project SAGE, decided to teach

System Thinking at the Sloan School at MIT, since all of this requires a transformation in behavior, and therefore a lot of explanations, training and mentors who are convinced and convincing; even with the advantages of the Web, it is a difficult lesson to teach, and a long-term one, since millions of European consumers will need to be convinced.

To go beyond this and put on a façade of demagogic willingness in order to please uninformed public opinions that no longer understand anything about the topic, or worse, industrial and/or legal lobbies, litigating under the umbrella of the principle of precaution (a real income for legal firms since it would be a never ending source of conflicts), this is the same as laughing in the face of the laws of nature and those of large-scale project engineering. As mentioned in Figure 7.5, complexity created in this manner risks going well beyond what engineers and the most experienced teams are reasonably able to organize and master. Technical governance cannot be subordinated to political power, for the risk of recreating Chernobyl is known. Chaos would be certain; all we would have to do is wait.

There is clearly an optimal system size beyond which it becomes difficult to venture. Beyond a certain threshold, sharing itself becomes problematic. Even with a good overall model, which currently does not exist, it is not clear that we would be able to implement it, given the mentalities and the diversity of actors: many rights and few obligations, unless we evolve into an authoritarian society. Only detailed simulations, still to be invented, will be able to provide us with some clarity (so to speak!).

Wanting to Europeanize the network and energy exchanges means, operationally speaking, establishing a system of systems governance created around a very large organizational perimeter with numerous interested parties, all with contradictory interests. It therefore means, automatically, creating organizational and human complexity, a problem that was already clearly identified by John von Neumann and Jay Forrester at the time of the SAGE Project, but a complexity that still remains misunderstood (see Part 2 of this book). As for the transport network, it is a supranational authority that cannot behave like a multinational company because of energy policies and strategic state interests. It would be naive, imprudent even, to imagine that all this could work without additional risks.

A first difficulty comes from the disparity of decision-making cycles tied, on the one hand, to the physics and engineering of the network, and, on the other hand, to the human decisions and political choices that can vary along with parliamentary majorities. On the systems side, it is a long-term affair, with heavy investments, whereas on the human end of things, it is the short term that prevails. A political decision, like that of the German Chancellor to abandon nuclear energy after the

Fukushima tragedy, is made in a few weeks, or at worst a few months, but the impact on the system and its technological ecosystem will develop over the course of decades, and after a certain threshold, without the possibility of going back, given how investment cycles work. We are beginning to see the first consequences of this in Germany.

A second difficulty comes from the psychological and cultural differences of actors. In a risky situation, an individualistic French person will not react as would a German who is used to collective work and a culture of compromise. In a multicultural context, there cannot be a "center" in the usual sense of the term. If the political situation fluctuates, national actors will remain in a state of uncertainty, and in the worst of cases, defiance and withdrawal will prevail. Any decision must be collective, which can pose a problem in situations where a decision must be made quickly. Risk management is a fundamental aspect of effective governance; here it is at its maximum intensity, with very little possibility of feedback for the moment. Collective work is an operational mode that major states are very familiar with when managing crises, for example, in NATO. To be efficient in this kind of situation requires ad hoc training, a common language such as the one defined by the *Military Standard Common Warfighting Symbology*, already cited, common working methods such as the ones defined by NATO in the MEDO/MARS guides, which are applications of sociodynamics to the world of military operations[18]. In summary, an entire set of learning and an essential level of qualification is needed to avoid causing catastrophes.

Finally, the third difficulty: there is a real problem with the skills required to work at this level of abstraction, so far from the realities "on the ground". The temptation to escape into the virtual world is a real risk that can make us forget the physical nature of the high-energy phenomena that the network carries, somewhat like the difference between macroeconomics and the real economy according to Schumpeter, all other things being equal. Making the decision to connect a high-power wind farm to the network is ipso facto ensuring that the power released into the network is compatible with the transport capacity, in particular for the reliable meshes; the one does not go without the other. However reliable the representations of the state of the network on the system control screens may be, we must never forget the physical reality. Behind a screen, the perception of risk is different; it is a well-known psychological phenomenon in high-risk jobs that can cause deviant behavior or reckless risk-taking. At the very least, proper training in *System Thinking* is vital for learning about the counter-intuitive phenomena caused by the interdependence and retroactions that go hand in hand with the system.

18 See: https://en.wikipedia.org/wiki/Social_dynamics.

Governance of such a vast ensemble is a formidable architectural feat. If roles are not well defined, anarchy is guaranteed over an energy volcano. Governance is the nerve center of the system, and in a certain way, a primordial aspect of its long-term memory. The operational symbiosis of the actors, in essence the team spirit, must be as perfect as possible, remaining aligned with the underlying physical reality which has its own, intangible laws. It must be organized so as to resist all sorts of lobbies that will stop at nothing to obtain advantages from this new decision center – advantages from their point of view. Each of the PESTEL (political, economic, social, technological, ecological, legal) factors will cause different types of pressures. In the case of conflicts – inevitable at this level – they will have to be refereed and the arbitrations will have to be made public in order to maintain the confidence of interested parties and, behind that, the confidence of users. The laws of physics mean that the interconnections of networks create a unique, very large physical object that can no longer be controlled as everyone likes, or else we risk catastrophes – similar to what would happen if a vehicle had more than one steering wheel. Cooperation and a sense of the common good become obligatory, but how do we guarantee this behavior as a matter of ethics? We are once more confronted with von Neumann's dilemma: "Who keeps watch over the watchmen?"

IMPORTANT.– All of this is easier to say than do since we still lack the necessary critical distance to judge appropriately, but there is no escape: we must be organized to cooperate, reflect, explain, act; but above all, we must not flee into the virtual world, however smart it seems to be!

7.6. Conclusion

The most important lesson of ES, from a systemics point of view, is without a doubt in the organization of levels of regulations and in the intimate knowledge of the system. We saw that risk is tied to the quantity of energy being regulated. The higher the quantity of energy, the more difficult it is to establish compensation within a continually shorter required amount of time, unless we keep reserved energy stocks on hand, in contradiction to the sought-out objective. Beyond a certain threshold, it must be split to provide reasonable energy "packets", a very small part of which will be used for compensation.

Concretely, and this is only an example, if 10 regions each leave 10% of their production quota for consumption, and never any more, to be regulated by a higher-level regulator, the latter gains a cumulative quota equivalent to what we now have, according to the reasoning suggested by Simon[19]. But for that to work, the principle of subsidiarity that we have just mentioned must be strictly respected by all

19 See *Sciences of the Artificial*, Chapter 8, already cited.

actors. Among the 10, none must drop their guard thinking that the superior authority will always be able to get by, and that inversely, the superior level should not get involved with that which does not concern it or for which it is not qualified; or worse, create compensation agreements without informing the others. Each must be locally free, to be sure, but above all, responsible. This rule of subsidiarity is, in fact, a variation of the Chinese doctrine of non-intervention, the *wu wei* of the Taoists, which is quite the opposite of laissez-faire; given that the Chinese have an ancestral political science, and a capacity to manage much larger groups of people than we are able to in Europe, perhaps there is a message of wisdom in there that we should also follow in our engineering practices. It is incidentally what layered architecture teaches us, letting us organize the millions of lines of code that our operating systems, our telecommunications protocols, and many more are made of. By a strange turn of events, it is in this context in the 1970s–1980s that we truly began to understand the virtues of modeling, thanks to the use of computer technology, when it was still in its adolescence, that let us explore models by simulation. With a higher degree of maturity, in the CMMI model, its quality as a kind of organization is no longer visible because it is everywhere, in all of our heads.

Many other aspects of systemics could be illustrated by electrical systems, such as the implementation of environmental considerations, that is, the *outside* of the ES. The ES is constrained by all kinds of factors which can be gathered under the acronym PESTEL, for the sake of practicality[20]. Each of these factors weighs on the outcomes of the ES. The production of electricity was at first a private affair. In 1945, it became a national interest. In the 2000s, following deregulation laws and at the behest of Europe, the EDF/GDF group was broken into many components. A management operator for the transport network was created: RTE (*Réseau de transport de l'énergie* or the Power Network), whose task was to ensure equitable service among all energy producers operating under the supervision of the CRE (*Commission de régulation de l'énergie* or the Commission for the Regulation of Energy), created in March 2000, which includes nine appointed members. The ecological factor now plays a major role. It has become almost impossible to create new infrastructures like very high voltage (VHV) lines, which make overall regulation possible, when the pressure to interconnect countries and regions is always growing, automatically causing instability because of these new couplings. The current infrastructures must be optimized using higher and higher performance simulations, made possible by the intensive calculations that are now possible. Nuclear energy has become a social interest that varies according to the political majority of the moment, or in cycles of four or five years, while the cycles of engineering, including the dismantling of the plants, along with the economic

20 See, in particular, Andrew Gillespie, *Foundation of Economics*, Oxford University Press, 2011, and the website (the entire article is interesting): http://www.kantakji.com/fiqh/files/env/ty3.pdf, with a PESTEL excerpt.

consequences, can be counted in decades (approximately 100 years for the EPRs that are currently being built). This is a new source of instability, and we will need to ensure that it does not become a hindrance to the quality of energy services or cause safety risks and waste. But whatever we may decide, we must never forget that in this gigantic energy machine, organized to the best of our knowledge and know-how, it is always nature that imposes its constraints, including the human contingencies. It remains a "small machine" at the scale of the Earth's energy machine and the devastating power that nature can release in hurricanes or earthquakes.

Here again, our only recourse for seeing clearly is the simulation of energy scenarios, subject to the validation of users, to calmly judge the situation. It is better to take the three to five years needed for this validation than to go down paths and obtain investments that might become dead-ends and that no one can guarantee today.

The complexity of systems like the ES creates "citizen" constraints, as we like to say nowadays, and even ethical ones. The operator of the system, helped by the regulator, must be able to explain "how it works" in terms that are understandable to a majority of consumers who, we can hope, have enough maturity to measure what is at stake, something that was not the case 20–30 years ago. Today, 80% of a given age bracket has a high school diploma! Information must be open so that independent organizations can obtain it and provide a contradictory opinion, in terms of risks, as is required in quality control. If we cease nuclear energy production, there are only two alternatives: (a) compensate using other forms of energy and/or (b) drastically reduce consumption, keeping in mind that demand will increase, even though only by demographics. Putting nuclear energy in a country that does not have PESTEL maturity (in the sense of competence, a properly validated qualification), what American economists sometimes call graduating, is a risk that would best be avoided[21]. Refusing to create production and/or interconnection infrastructures means accepting, for the common good, to voluntarily reduce one's needs and/or endure power outages.

Illustrated in an ad hoc way, with simple pedagogical models that computer technology now makes it realistically possible to create, systemics can help the citizen, the political decision-maker, or the locally elected official to better understand the complex world that is now our own, to enlighten their sense of discernment and their sense of responsibility. It is better to bet on collective intelligence over ideology, a lesson we have already experienced with the repeated ravages of the 20th century.

---

21 See the works by Reinhart and Rogoff, and their most recent book, *This Time Is Different – Eight Centuries of Financial Folly* (Pearson).

POST SCRIPT.– We cannot resist drawing a parallel between the electrical system and the financial system, the latter of which was progressively established after the first deregulation measures, beginning with the non-convertibility of the dollar, taken by President Nixon in the 1970s.

REMARK 7.5.– Anecdotally, let us recall that in the case of the ENRON scandal in the United States, the crooked financiers at the head of the company had managed to sell energy that did not exist with the complicity of Arthur Andersen's consulting firm, which was falsifying accounts. Californian users paid the cost by being subjected to many power cuts. After these directors went to prison, everything went back to order. Arthur Andersen disappeared from the scene, discredited by the scandal.

We cannot play poker with the laws of physics as we can in an unregulated financial world, with impunity. As for the socio-economic consequences of the crisis caused by Covid-19, it is a tidal wave that will wash over the planet and that only, given the state of our knowledge, systems will allow us to understand, as mentioned in the conclusion of Chapter 1.

8

Systemic Aspects of Project Systems

Systemic features of projects have long been emphasized, in particular by Gerald Weinberg, in his four volumes of *Quality Software Management*, especially in Volume 1, subtitled *Systems Thinking*, which applies to all types of systems. The MERISE method, which had its heyday in the 1970s–1980s, was also based on systemics. We will only cover the general aspects that are particularly illustrative of systemics here.

This chapter is also the fruit of many years of experience teaching project management for the engineering school at the CNAM, where I was co-director during the 2000s, and the experience I acquired in the actual management of the large projects I was in charge of, which served as a foundation for my teaching.

8.1. The science of projects

This is not the place to present the "science" of projects. For that, there are excellent books, paired with normative documents and guides, such as the ones produced by the Project Management Institute[1]. What we wish to show in this chapter is the way in which project systems and the quality control component fit perfectly within the phenomenological laws of the *Compendium*, while incidentally revealing the relevance of complexity analysis as we propose it.

In her book, *La gloire des ingénieurs – L'intelligence technique du XVIe au XVIIIe siècle*, the philosopher and historian Hélène Vérin considers the "estimate [to be]

For a color version of all the figures in this chapter, see www.iste.co.uk/printz/pedagogy.zip.

1 The main part of my academic work has been the exploration of this dimension of information science. For guides, see the site: http://www.pmi.org, and for the downloadable version the guide: http://www.cs.bilkent.edu.tr/~cagatay/cs413/PMBOK.pdf.

the masterpiece of the engineer", the criterion for the maturity of well-assimilated engineering, and rightly so; for, how could we establish a reasonable estimate without knowing all of the aspects of the thing being made[2]? We should note, incidentally, that this science of the estimate is a compromise between the approximate but sufficiently precise knowledge that the physics of the materials we use provides us with – the sciences of the artificial, to use Simon's term, with information making it more complicated since what those projects work with is, in essence, immaterial – and the capacity of engineering teams, in the human sense of the term, with their specific contingencies, to employ their knowledge for concrete creations. To build technical, finalized objects/systems, human creativity is essential. Simondon's triplet, {U/user, S/system, E/engineering} allows us to ideally situate the place and time of this human component that makes it possible for ideas to lead to concrete objects/systems[3]. The systemic approach in turn lets us explore all of its aspects at the desired level of abstraction, including the auto-adaptive aspect.

The place where this science of the estimate gets used is the engineering project, specifically, of the size of the project in terms of the human resources needed, both in quantity and in quality. It is the process by which an idea takes form and is materialized as a system. As Bachelard predicted in his 1934 book, *Le nouvel esprit scientifique*, where he said "above the *subject*, beyond the immediate *object*, modern sciences are based on the project" (introduction, p. 15), the project became the keystone of information sciences and their materialization in the constitutive systems of cyberspace. This science of projects, including the human component known as social dynamics, is now an essential aspect of engineering that grew from the SAGE and NTDS projects.

DEFINITION 8.1.– *By projects science, we mean the set of human means and the methods that we create for ourselves to transform an initial situation into a new, final situation that is more favorable to us overall. It is a science that teaches us to navigate complexity by resolving the problems encountered while avoiding the insignificance created by a combinatorial.*

2 Published with Albin Michel, 1993; to which we could add the book of homages, *Penser la technique autrement*, Classiques Garnier, 2017.

3 A technical object/system technique like the Airbus A350 is the result of the work of about 60,000 people, eight years of R&D, hundreds of sub-contractors, five devices covered in monitors for testing and providing the system with industrial information, for about 10 million euros of investments; the first commercial flight took place in 2015 (source: Airbus Industry; http://www.airbus.com/aircraft/support-services/publications.html). My colleagues at the Institute of the Economy speak of a "work brain".

In information engineering, it is impossible to eliminate the human factor, an intrinsic component of complexity since information is a human creation; it is constructed by humans and for humans, even if computers happen to be interposed. This is a radical difference from the classical engineering of engineers, where everything is done to eliminate the human factor from the laws of physics. The laws of physics and/or chemistry are "objective", and the "subject" who develops them and uses them has no place in it. In communication, in Shannon's sense, anything that processes meaning and interpretation has been voluntarily excluded in order to conserve the objective character of technologies used to communicate, that is, principally, the materials that conduce electricity and/or electromagnetic waves, the communication "channel" in Shannon's theory.

With information, it is the exact inverse. A good programming language in the widest sense of the term, such as the modeling languages SADT, IDEF, UML, SysML, BPMN/BPEL, etc., all used in systems engineering must be "context free", or independent of the speaker – in other words, the person doing the programming – to avoid any ambiguity and/or subjectivity that may make it incomprehensible to a third party. It is the organization, the structure of the information and the relationships that will carry the meaning of the object/system being built. Despite certain similarities with human languages, such as the notions of syntax, grammar and pragmatics, the difference is profound. The objectives are not the same. Human language easily carries emotions, nuances, all sorts of stylistic and/or metaphorical figures like the oxymorons that must be excluded from programming languages, which are more like cold monsters!

Natural languages and artificial languages have, however, from the point of view of users, a profound connection of primary importance: the errors made by speakers. With a natural language, mankind adapts more or less to all kinds of errors. Our brains and our cognitive capacities are very tolerant with respect to errors and disorder, in which we often seem to be at ease thanks to semantics, the objective of our actions and "common sense". It is probably a capacity for survival developed by our species over the course of evolution. With computers, the situation is reversed since they are intolerant to the slightest error. One simple error in programming, one mis-programmed instruction, and we have the Ariane 501 rocket exploding a few seconds after liftoff.

For a human operator in "normal" conditions, the average rate of errors, that is, erroneous or inappropriate actions, is on the order of 5–10 errors per hour of effective activity. Most of these errors get caught, but given their number, there will always be a residue of errors that slip through the cracks. We know there is a relationship of the kind $Effort = k \times (Size)^{a}$, with $\alpha > 1$, which expresses the nonlinearity of human effort with respect to the size of the text that organizes everyone's work. With software, we can thus have a template of statistical averages, or, in hours of work (mh, or man-hours), 2.5 lines per hour per engineer, observed over thousands of projects and millions of lines of code. For software containing a million lines written by programmers, it takes at least 400,000 hours of work; this suggests the breadth of the error phenomenon. The rate increases greatly in cases of stress, and is variable depending on individuals, on the state of fatigue and all sorts of emotional factors. To diminish the error rate, as few as possible must be made at the source; this means reasoning correctly, given that any program is first and foremost the result of reasoning, and avoiding doing more than one thing at a time, which perturbs our brain… but we can never be sure to eliminate them entirely.

Once again, we find, at the center of our considerations, three aspects: 1) the functional form of software and control over the size of objects; 2) interactions/exchanges, that is, communication and interoperability, and 3) system safety, error discovery, monitoring and repairing, that is, quality assurance in the most general sense.

8.2. Control

8.2.1. *Project size*

If we follow the logic of Simondon's triplet, {U, S, E}, at the most basic level, we are faced with three sets, as indicated in Figure 8.1.

In the case of the Airbus A350 that we mentioned, set U includes all the airline companies that had an A350, and by transitivity, all their passengers, that is, millions of people. Set I represents all the people involved in the A350 project, which, according to Airbus sources, was about 60,000 people, mostly engineers, both employees and consultants working for sub-contracting companies that contributed to first-level airplane development and maintenance. Set S is made of all the material pieces integrated in the airplane and the ones in the machines that contributed to its manufacture, called the *bill of materials*, which gave rise to production management, or *manufacturing resource planning*, now largely computerized[4].

To this material subset, we must now add a subset of immaterial elements, that is, the lines of code integrated in the computers that make up the nerve center of all our systems today, let us say since the 1990s, when the shift towards having

4 See: https://en.wikipedia.org/wiki/Bill_of_materials and https://en.wikipedia.org/wiki/Manufacturing_resource_planning.

everything digital became irreversible (see Figure 8.8). It was a revolution without precedent since the material components were elements that on their own did not require much attention as to the atomic/molecular constitution, which is not the case for the immaterial components and the computers that sustain it, made of nanometric components that are subject to the hazards of the quantum sphere.

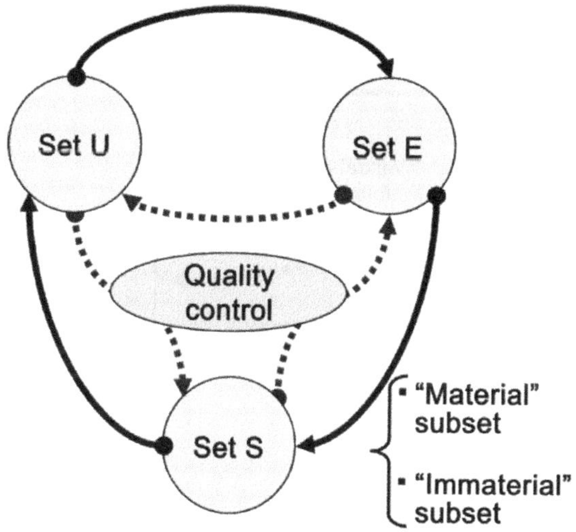

Figure 8.1. *Sets involved*

We are told that in the near future, there will be self-driving vehicles that will integrate tens of millions of lines of code and about a hundred computers that are themselves connected to data centers containing hundreds of thousands more[5]. Here is the source of the combinatorial challenge that we will have to confront.

The great novelty with the mass computerization of all systems is that the information that has always come with the system, even if just the documentation containing user guides, is no longer only passive; it is now active, and at the very heart of the system. A faulty instruction, and we have the destruction of the Ariane 5 rocket launcher on the first attempt. It is therefore important to pay very close attention to any information that plays a role in the system's overall programming. This active information is generated by the engineering cycle shown in Figure 8.2. For the system to work correctly, it must be integrated, not only with the computers that give it "life", but also with the rest of the system that makes up its environment.

5 Certain builders have spoken to us of 100 million lines of code that will need to be validated!

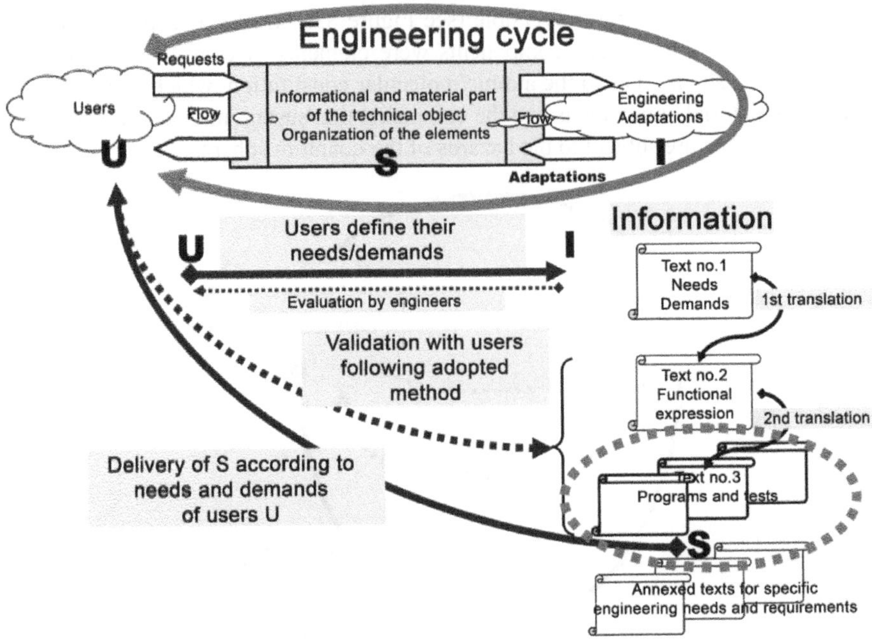

Figure 8.2. *Active information*

What is interesting in the case of the Ariane 501 is that there was certainly an inadequate instruction, but the fault should have been detected by the integration tests if these had been carried out correctly, therefore before the fatal launch. The error in programming, already present in the Ariane 4 navigation system, had never become a fault, and this was the case in the last 40 launches of Ariane 4, but the trajectory of Ariane 5, a much more powerful launcher, was different; the environment was thus different, whence the fault, since if an Ariane 4 were on the trajectory of an Ariane 5, it would mean it was falling, which would lead to the inevitable destruction order.

The lesson of this story is that active information obviously refers to the execution of the code, but it also includes the tests.

This immaterial dimension was perceived from the start, in particular by Jay Forrester, the director of the SAGE project, who pronounced himself very clearly on the subject[6]. Starting with this foundational project, the quality aspect began to be

6 See his article, "The impact of feedback control concept on the management sciences," from 1960, republished in the volume *Collected Papers of J.W. Forrester*, MIT Press. It is incontrovertible!

taken truly seriously, since within the principal engineering cycle that takes place in the direction of U → E → S → U → etc., there are secondary cycles generated by the relations between U, S and E:

– between U and E, to make sure the needs of users are correctly defined and will be able to give rise to contractual requirements that the system's cost estimates will be based on;

– between E and S, since as with all technical objects as long as engineering has existed, the system's material components are progressively constructed by steps where, starting with an initial seedling S_0, a series of incomplete versions will be derived $S_0 \rightarrow S_1 \rightarrow S_2 \rightarrow \ldots \rightarrow S_{Nominal}$, with intermediary IVVT, until the final version expected by users is reached;

– between S and U, where, as soon as an intermediary step S_i can be reasonably presented to a subset of selected users, it is best to close the cycle to make sure that what has been made by community E indeed corresponds to the expectations of users U. These are the ABCs of quality control.

The primary role of quality control – an essentially human system – is therefore to make sure that all these retroactive cycles work properly on the basis of validated information. This typically falls within the purview of empirical models, the *theory-less* of L. Valiant (already cited, see Chapter 3 and the references), which must nevertheless build upon uncontestable objective observations. Parallel to the flow of information that accompanies the progressive construction of the system, there is an inverse flow that goes upstream from the point of information, which in the jargon of quality control models gets called *findings*, which can be a) avowed errors that need to be corrected or b) desirable improvements that will be useful or even vital and whose cost will need to be estimated to remain within the initial pricing.

The quality control system therefore has a controlling function of utmost importance and must be given the proportionate weight, its role is not to replace those who have not done their work or have done it badly, but to make sure the interactions of groups {U, S, E} are fluid and are addressed in the proper order. It must ensure that all the retroactive cycles are indeed closed and that the inverse flows are taken into account so that the transformation process is maintained in equilibrium as it unfolds.

The quality control system also has a redundancy function tied to the hazards of the real world. If the world were perfect, if there were no such things as erosion or degeneration in the wider sense of the term, if there were no errors and therefore no faults, there would be no need for quality control or a quality system, which would be reduced to the bare minimum. This is what the maturity models that took shape in the 1990s worked towards, under the auspices of the United States DoD and the SEI[7]. Here again, qualitative empirical models were used, but were nevertheless very useful.

8.2.2. *The dynamics of relations: pairings*

We saw in Part 2 that alongside basic numbering, the relationships that develop within sets and between sets must be looked at. If we take the case of A350 with its 60,000 people involved in the project, in other words in community E, we cannot imagine a pairing matrix of 60,000 × 60,000, which would be absurd. The organization will necessarily be by projects, starting with relatively autonomous teams that must not, as has been known for some time, surpass 10 or so members; this is the famous rule of "the magical number 7 plus or minus 2", known by all project managers. In the 1960s–1970s, the economist François Perroux, a professor at the Collège de France, had theorized it with what he called "active units", or a small group of people in a symbiotic relationship with their resources so they had a certain level of autonomy and had to report to a superior authority[8]. It is a notion that is also found at the heart of what has, since the 2000s, been called agile management and whose purpose is to fluidify action, the real question being "how?". With teams of 10 people, we still have a matrix of 6,000 × 6,000, which continues to be unmanageable.

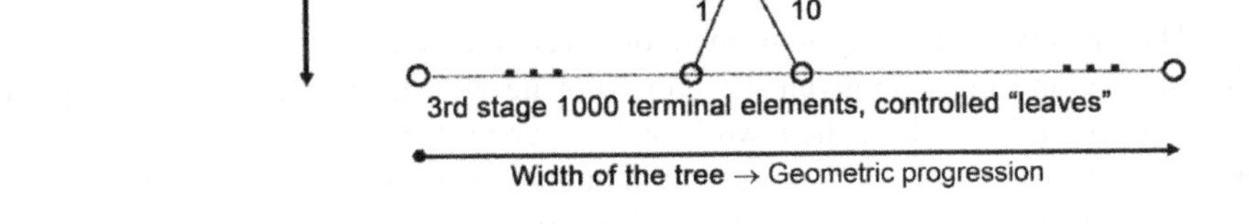

Figure 8.3. *Organization control*

If we put the 7 ± 2 rule into practice from the base of a hierarchical organization, rounded to 10 for the simplicity of the demonstration, we have a capacity of control

7 See: https://en.wikipedia.org/wiki/Capability_Maturity_Model_Integration.

8 See Perroux, F. (1975). *Unités actives et mathématiques nouvelles*. Dunod, Paris.

of 1,000 terminal elements in two stages, or 111 control elements (approximately 10%), but with four stages we get 10,000 that we can organize into 100 big packets of 1,000, which will in turn be managed as a system of interoperable systems according to what was recalled in Chapter 1.

This mode of organization can only be envisaged if the information flow within and between systems is rigorously managed, and identified in the first place, using a reference framework that we will discuss.

8.2.3. *Human complexity*

On the user end, the situation is different depending on the nature of the objects/systems being made. Either they are objects/systems for restricted communities like the LHC at CERN, or the EDF/RTE regulators of the energy transport network, or even those for air traffic control; or they are for the general public, like users of automobiles or smartphones. In the case of the general public, the population in question numbers in the millions, even billions. In any case, complexity is everywhere.

To avoid the combinatorial that a direct relationship between the "engineering population" and the "user population" would give rise to, an intermediary interface that will channel interactions is needed, as shown in Figure 8.4.

The left side of the figure is an abstraction of what happens in the world of users and the processes that they make use of. We could complicate this much more by, on the side of the general public, including the networks of distributors country by country, since depending on cultures, sales and distribution are different. We do not speak to someone from China as they would to someone from France or America. On the engineering end, it is different since the object/system being built imposes its own logic, hence the relative conversion of good practices on standards that are today generally accepted no matter the country or the culture[9].

The quality control system that was developed in the 1950s with people like Juran, Deming or Crosby, all Americans, was first rolled out on a large scale in the Japanese industry, which made a secret weapon of it with the success that we are now all aware of, returning to the United States and Europe some decades later with the designation of lean management.

9 Such as the INCOSE Systems engineering handbook, or the ISO/IEC 15504 and 15288 standards that define engineering processes.

8.3. Volume of information exchanges in projects

8.3.1. *Interoperability {U, S, E}*

We should have understood by now, at this stage, that a) the exchange of information between the components of the triplet {U, S, E} is continuous and intense, so as to maintain the overall coherence of the system and ensure its proper functioning, and that b) errors will automatically metastasize if they are not detected and corrected. Each of the components has a reference specific to it, but for the ensemble to work, each of them must translate into the language of the two others through a natural process called transduction, which our intelligence works to perfect; transduction was succinctly presented in Chapter 2.

Nature abounds with natural transduction mechanisms that have not ceased to attract the attention of scientists and/or philosophers since time immemorial, for example, the learned Greeks from the Milesian School in the 7th Century AC. The mathematicians who discovered logarithms showed a strange correspondence between the operation of multiplication × and the operation of addition +, which is another way of projecting a surface or a volume on a line. They immediately put this to good use. The affair was only truly elucidated with the discovery of the theory of groups in the 19th century, then algebraic structures, rings and bodies, and finally the generalization of the abstract correspondences of algebraic geometry by Grothendieck and his theory of schemes/motifs. As Hilbert had already pointed out in his *Foundation of Geometry* (1899), the foundational entities of any geometry (point, line, surface) determines a relational structure that will give rise to different geometries, with the variants of the axiom of parallels; but the most remarkable part is that there is no order of presence between these three entities, and they are mutually definable in terms of each other. A point is the intersection of two lines, or perhaps a tangent as in the theory of envelope curves where, now, we have a correspondence between points ↔ right angles. To describe and visualize our familiar geometric space, we need all three entities simultaneously and in a timeless way.

Whence our remarkable ternary structure, progressively exposed with the help of information sciences, this science of the artificial, to paraphrase the title of that seminal book by Simon. The use cases formally identified in community U, a series of actions considered relevant in other words, will be transformed and generalized into a functional continuum, the heart of the exchange reference framework R_E, materialized in an organic structure S through the transformative action – a transduction in the strict sense of the term – of engineering processes implemented by community E (see Figure 8.6).

This is the place to recall the fundamental distinction between communication and information, a distinction recalled repeatedly by Shannon himself, so much so that confusion reigned following his famous dissertation, *The Mathematical Theory of Communication*, from 1948.

Communication deals with the unaltered transmission of an energetic signal between an emitter and a receiver; it is fundamentally a syntax problem whose coding is only one aspect, in conformity with the underlying physical phenomenon.

As for the information, it relates to signification, meaning, and is carried by the signal. For there to be information, there must be a device for interpretation that will trigger actions, activate processes, etc. For that, the actors sending and receiving messages must agree to conventions among themselves regarding the meanings attached to the messages qualified as meaningful, following a distinction made in the first books on linguistics, and organize the composition of signifiers with grammatical rules such as those of Indo-European languages[12]. The logician Wittgenstein masterfully analyzed what he called "language games" in two books, the *Tractacus logico-philosophicus*, which dates from the 1920s, and the *Philosophical investigations*, from the 1950s, in which he takes up and expands on the *Tractacus* during the period where information sciences were appearing. One of the leitmotivs in Wittgenstein was to get students to pay attention to "how it is used!", one of the lapidary formulas in *Tractacus* saying, in Aphorism 3.328: "If a sign is not necessary then it is meaningless." In information sciences, if we do not understand the purpose of the signal that has been received, then this signal is simply "noise" that we cannot do anything with.

Information, for the most part, is independent of the means of transmission, but it needs some kind of support to be materialized; the support is therefore neutral albeit essential. A verbal message, and therefore a sound wave, will have exactly the same effect as a written message on a material medium, even if the information is the same, that is, a transduction in terms of actions to be taken.

This is where our ternary structure {U, S, E} takes an entirely different dimension, since it must make it possible for information to circulate between the three components independently of any communicational medium. It is the existence of this timeless invariant materialized by the languages used that creates the possibility of implementing a system both useful to users and realizable by engineering teams (see Figure 8.7).

12 See *Cours de linguistique générale* by Ferdinand de Saussure.

Each group/community has a specific reference, but the three have an exchange reference in common, an exchange model that guarantees that what gets expressed in the language of U will be translated into the two others, and vice versa (see Figures 8.7 and 8.8).

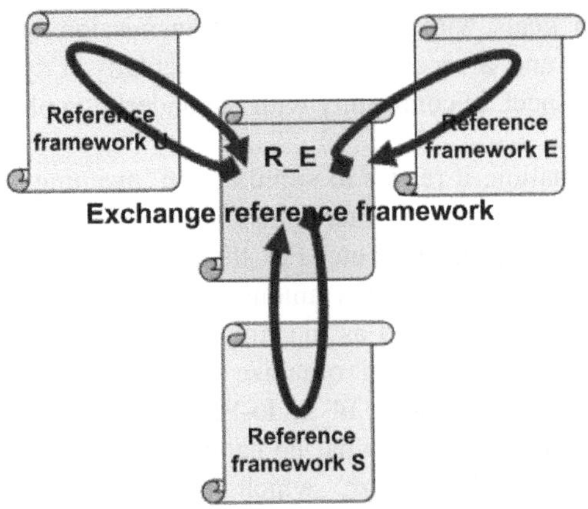

Figure 8.6. *The exchange reference between {U, S, E}*

In other words, what is specific to U is its real environment, that is, in the world that we all have in common, our common property, will be representable in an abstract functional world that is strictly equivalent – a world that, once transformed and informed with organic elements, both material and immaterial, and properly arranged, allowing the system to be physically constructed and bringing it into existence, will allow U to act more efficiently than it could without the system, and even do things that would seem impossible to users, such as flying.

It is the existence of this constant that will allow for the integration of organic elements and the construction of a new, finalized object/system, endowed with properties that did not exist in any of its parts. Integration causes the emergence of a new being, with a logic that is an acceptable and formally accepted compromise between the three components {U, S, E}, endowed with a specific individuality. The process of integration now appears to be the fundamental operator of the emergence of new things. Figure 8.8 shows the complementarity of the organic hierarchy and the informational hierarchy, two indissociable, however, distinct structures.

If one of the retroactive cycles in Figure 8.6 breaks for any reason, if the languages become incoherent, the integrated referential entity {U, S, E} loses its specificity, with the consequence being a progressive disintegration of the relational structure that makes the three groups {U, S, E} into one integrated unit: a monad. The essential thing to understand is that this "reference framework exchange structure" R_E is a dynamic structure that only lives and survives through the permanent exchanges that exist between its components.

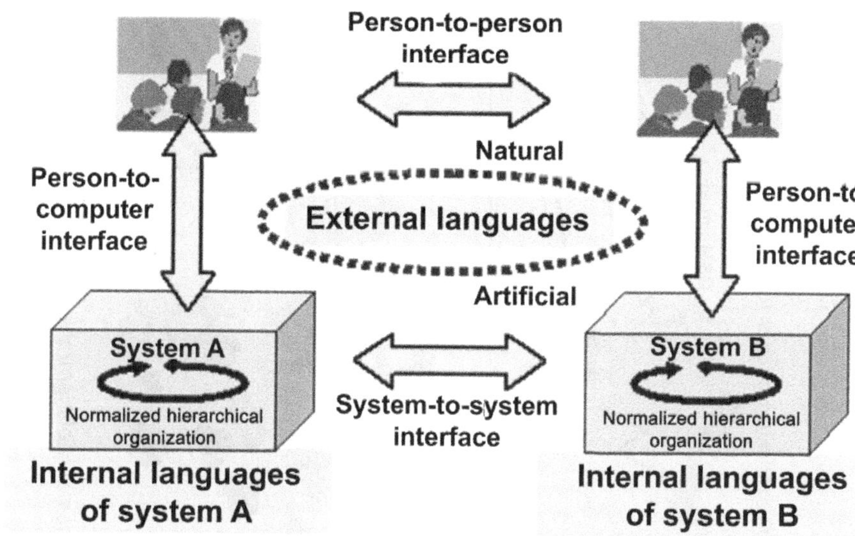

Figure 8.7. *The languages of the exchange reference framework*

A reference that does not get used quickly becomes a dead language, like the languages that get referred to by this name. If the human energy – in quality and in quantity – that underpins it disappears or reaches critical levels, the reaction goes out like an autocatalytic reaction that has not been maintained. To understand the effects of this, Figures 8.9 and 8.10 are essential.

What makes the system useful is that the energy spent bringing it into existence, especially when using the {U, S, E} reference, "frees" energy with regard to its users who, thanks to this lever, will be able to do something else, something more than what they would previously do. After a certain amount of useful life, the gain users make compensates for the cost of the engineering, the moment which gets called the break even point in the jargon, the equilibrium point in a project without which no project would be profitable. The analogy with what gets called "free energy" in thermodynamics is evident here, but we will not develop it further.

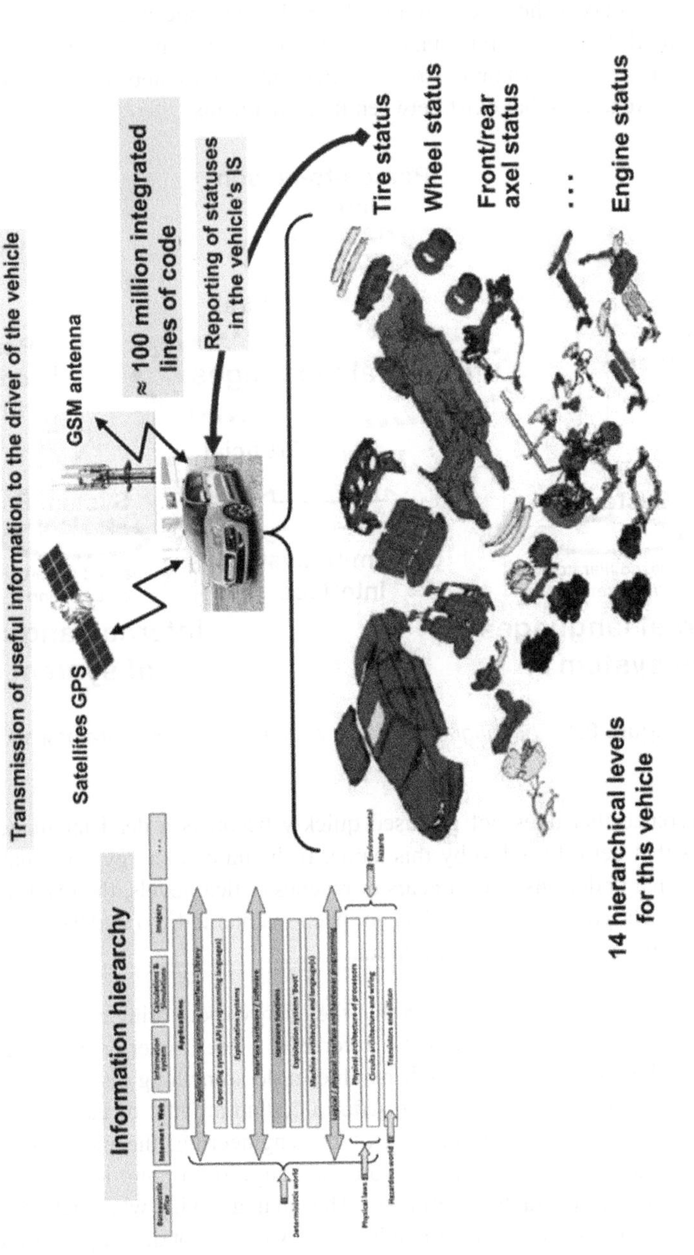

Figure 8.8. *Complementarity of hierarchies*

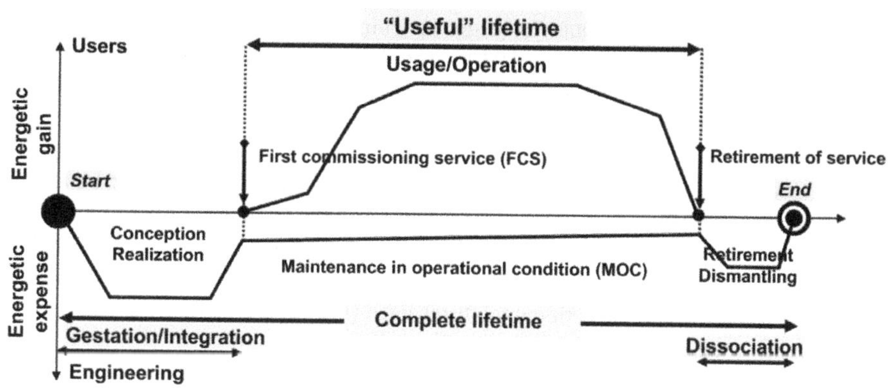

Figure 8.9. *The nominal "life" of a system {U, S, E}*

Airplanes such as the A350 or the Dreamliner-787 from Boeing get studied for having survived for about 40 years, or three generations of engineers; the latest generation of nuclear power plants will last for at least 100 years, or five generations, during which the reference will have to survive to ensure it functions properly and to feed the evolution of the reference itself.

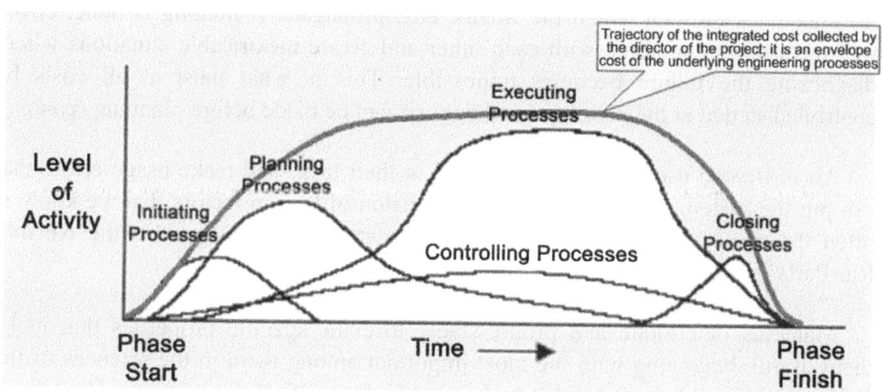

Figure 8.10. *Analytic structure of conception/realization costs*

To make a metaphor, we could say that human society has mandated some of its members who are aptly qualified to make a system that will free energy that can be recovered by the greater population. If we take the example of the Airbus A350, we have, on the one hand, the project with its 60,000 contributors, and, on the other hand, the hundreds of millions of passengers transported throughout the life of the A350. At the time of the first contact with the Far East, it would take at least six months for the Jesuit missionaries to get to China or Japan via the maritime route opened by Vasco de Gama; in the 1930s, it took three to four weeks by boat; today, we can complete a Paris-Peking or Shanghai route in 11 or 12 hours in one flight. We can measure the gain.

However, to understand how energy is freed from the project for the benefit of users, we must consider two levels of observation – the horizon of overall costs: the macro-level of patrons, in other words, the CQFD (cost, quality, functionality, delays), seen by users, and the intermediary meso-level, seen by engineers, process by process, depending on the nomenclatures used (see Figure 8.10).

8.3.2. *Detecting, monitoring and fixing errors*

Whether users and/or the engineering are communities of human actors or elements of the system, material or immaterial, errors are everywhere. Human actors make mistakes, and everyone works with a personal rate of errors in a particular context of action. Errors cause faults that can degenerate into variably serious failures. The materials and waves that constitute the physical part of the system also have faults since they are subjected to the hazards of the environment and aging, depending on the laws of deterioration that themselves depend on the materials used, the substrates through which the waves, etc., propagate. If nothing is done, errors will metastasize, combine with each other and create inextricable situations where diagnosing the failure becomes impossible. This is what must at all costs be controlled so that at the very least, a diagnosis can be made before planning repairs.

An ill-trained user, stressed, distracted in their task, will make usage errors that can put the system in peril; this is the dimension of human factors that we know is often the cause of accidents and sometimes catastrophes, such as in the AF 447 Rio-Paris flight on June 1, 2009.

Materials deteriorate and progressively lose the specific properties that make them useful, beginning with the most important among them in the sciences of the artificial: the molded silicon that makes up the heart of all our systems. The silicon in our computers is part of what the physicist Pierre-Gilles de Gennes called fragile objects; a nothing, a micro-energy, can make them lose their specificity and melt away their ability to carry information.

In nature, we can observe the existence of natural information, for which DNA and the cellular machinery are the carriers – in particular mitochondria; they were discovered in the 1950s along with a chemical alphabet made up of the four bases A,

T, G, C, texts that we now know how to sequence, but of which we understand but a very small part[13].

To this natural information, we have added artificial information, something which we have produced, with its flagship device: the computer, which builds on theories developed over the course of our history, some of which are already ancient, for example, mathematics, a universal language; these are the theories of World 3, from the philosopher Karl Popper. With information, we are in the infinitely complex, for which we already reviewed the rules in Part 2.

We will not speak of materials here, an ancient science identified by Galileo that continues to develop, but we will concentrate on what is specific to the science of the artificial: information, that is, the texts that materialize[14]. These texts, like any object in our physical world, contain errors due to the errors of the humans that produced them. The material support of these texts can deteriorate, so we add faults of a different order, including when the text is transmitted by any physical means. Finally, the readers/receivers of these texts can make interpretation errors of which they themselves are the cause, individually and/or collectively, or perhaps because the text is ambiguous, or a mix of the two.

Each of the groups {U, S, E} is a producer of errors. By observing Figure 8.9, we clearly see the overall issue.

8.3.2.1. *User errors*

Users, during the useful "life" of the system, will create a flow of errors that essentially depends on their level of activity with respect to the system. This activity can be measured in a) hours/minutes/seconds of connection, which the system itself can keep track of, including robots that can be connected and that work at a much higher cadence; or b) the number of requests made by users, robots included, which makes it possible to connect potential anomalies to the requests themselves, which is, semantically speaking, more useful, although to have a veritable measure in the sense of measurement theory, a representative "yardstick" request needs to be defined.

8.3.2.2. *Engineering errors*

Engineering and its thousands of developers will cause a flow of errors introduced primarily during the period of conception/realization of the system, and

13 As a reminder, the DNA of human chromosomes totals 3.4 billion bases ATCG, or about 1,000 volumes in the Library of America format; see Part 2.

14 See, for example, Guyon, E. et al. (2014). *La matière en désordre*. CNRS Editions, Paris; Guyon, E., Troadec, J.-P. (1994). *Du sac de billes au tas de sable*. Odile Jacob, Paris; or even Bak, P. (1996). *How Nature Works*. Springer, Berlin.

then marginally in all of the MCO activities that involve maintenance, and therefore modifications of the system, by teams. This means that when the system receives maintenance, the FCS (First Customer Shipment) in Figure 8.9, the system is delivered with a number of uncorrected, residual errors that will cause faults. The variety of possible situations means that the faults will become apparent or will remain latent during the useful "life" of the system.

The faults that appear can be placed into two categories:

– reproducible errors by the engineering teams in charge of maintenance in reconstituted environments that make diagnosis and then correction possible;

– non-reproducible errors, noted by users and/or clients but that the engineering teams will be unable to reconstitute for various reasons: lack of information, impossibility of recreating the context, etc.

In certain cases, the instruments introduced to observe what is happening in a system modify the temporality and synchronization of the underlying processes, with the effect of making the context in which the error appeared disappear, somewhat as with the Heisenberg uncertainty principle. We know that an error lost in the combinatorial of possible states exists, but the path to get there is inaccessible. The corresponding errors are called random, which is not necessarily crippling since if they are truly random, it suffices to restart the system from a coherent state; the new context will be different from the previous one, by definition, and the execution will then proceed without fault. Such is the magic of randomness, but for it to manifest itself, we must be sure that the fault is truly random.

It is a mechanism that gets used in fault-tolerant systems. In such systems, the designer speculates on the comparative latencies between the retroactive cycles of users – a bit less than one second for well-trained human operators – and that of the reparative mechanisms. If the time required for a repair is inferior to the time of the user cycle, then the formula for availability $D = \frac{MTTF}{MTTF + MTTR}$; we can write $MTTR = 0$; the perceived availability is therefore 100%. The latency of the cycle appears as inertia here, in the mechanical sense of the term; it is, in a way, its inertial "mass".

The only condition for all of this to work is having a detailed knowledge of retroactive cycles, according to the principles of systemics and the statistics for the occurrence of errors[15]. It is an interesting case of processing the ignorance we spoke of in Part 2 in terms of an empirical model.

15 See Printz (2019).

In conclusion, the frequency with which errors/faults appear throughout the life of the system, notwithstanding the activity that must be normalized and the quality of the engineering, tends towards the rate of residual errors that appears thus as a constant, which makes it possible to say if the system is acceptable or not in its context of use and of risks taken and accepted by users.

The detection of this invariant comes from the detailed knowledge of retroactive cycles.

8.3.2.3. *Errors in physical equipment*

These errors can be blamed on the materials and the physical phenomena occurring in the organic component of the equipment. These are errors/faults that engineers have been aware of for as long as engineering has existed.

The new feature introduced by information science is that any piece of equipment can be mirrored by virtual equipment, completely computerized, providing a complete image of it (see Figure 8.8). Here again, we have an empirical model of the equipment to monitor that, with the help of sensors/effectors – these are transducers – make it possible to interact with the equipment from a control center (see Figure 8.11). This whole ensemble can be teleoperated without any difficulty except for the means of communication.

The computerized virtual image is a reflection of the equipment's complexity, which must be taken into account to ensure the service contract for the system as a whole. The monitoring equipment will generate a flow of information that will be analyzed by centralized system management, which will need to be sized according to how much information is sent to it. Here again, we see that a detailed knowledge of cycles is a necessity, as well as statistical knowledge about the errors that may occur.

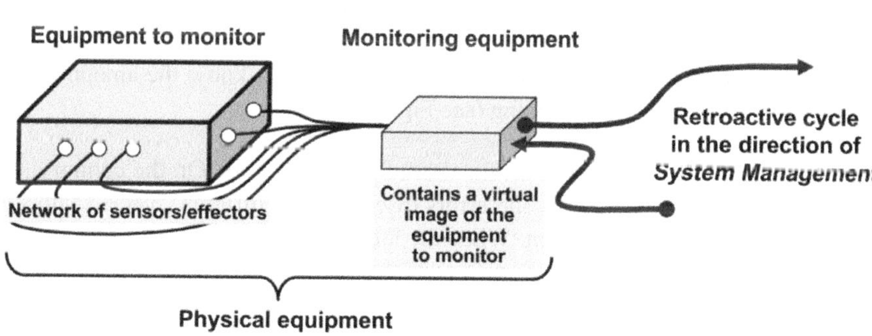

Figure 8.11. *Monitoring physical equipment*

The technical objects/systems in critical domains such as nuclear, space and aeronautical engineering already rely greatly on this type of mechanism to guarantee the service contract, for which a systemic analysis lets us estimate costs. This is the case with self-driving vehicles and smart objects in contact with the general public.

8.3.2.4. *Mechanisms for detecting errors*

These errors essentially concern the texts that are written by project actors, as seen in Figures 8.2, 8.4 and 8.5.

The first principle of systemics stipulates that nothing should ever be done without knowing how to control it, as recalled in Chapter 1. This turns out to be especially true when applied to the process of error discovery and correction. The integration process seen in Figure 8.5 validates that what was built by development indeed satisfies the demands of users.

The elementary pieces made by basic project cells – the activity units – enter the integration process with their validation tests according to the rules of the reference framework. In the case of the A350, these are hundreds of thousands of pieces, and in the case of computerized systems, there are several billion transistors in the latest generation VLSI SOC circuits, to which we must add the millions of lines of active code that will process the system's information. Testability, designing to test, is the key to all these systems. The integration process is the inverse image, the symmetrical image of the design process from which it is indissociable.

The question is therefore: how many pieces can we combine into one operation so that the errors discovered in the process can still a) be diagnosed, that is, the corresponding faults can be localized, and then b) corrected, by going back to the initial cause?

What must be controlled in this process is the effort required – this is measured in energy, which is a cost – to diagnose and correct the faults, given that their existence is given in development but for which we do not know the amount when the process of integration is beginning (see Figure 8.12).

Each piece contains an unknown number of residual errors. On the contrary, we have all of the information about the errors discovered in different pieces thanks to the validation run on each of them. When the integration process of *n* pieces begins, the validation activity will produce new tests that will lead to the discovery of new errors in the integrated pieces. The rate of the newly integrated pieces is reduced to an acceptable value for the subsequent operations.

The logical structure of the integration operator, denoted $\oplus$, can be given in the form of a so-called idempotent operation in algebra: $1 \oplus 1 \oplus 1 \cdots \oplus 1 = 1$.

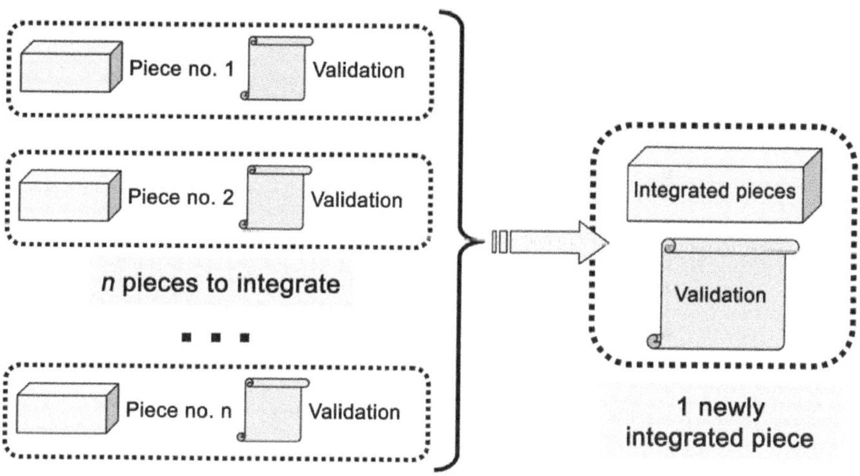

Figure 8.12. *Mechanics of the integration operator*

Two problems are thus left to resolve:

– choosing the pieces;

– determining the number of pieces n to process simultaneously in order to discover new errors.

The choice of pieces is made using pairing matrices, whose importance is well-known in systems engineering. All the modeling tools of systems allow for the construction and management of these matrices using *matrix modelers*. We will only provide the principle here (see Figure 8.13).

Two pieces are paired, here c and g, which are represented in the matrix by crosses at the intersections of lines and columns c and g; the presence of two crosses means that the pairing is in both senses, from c to g and vice versa. If the pieces are small, the size of the matrix can be large, very large even, several million with SOC transistors and/or lines of code, but not very dense – we say in this case that the matrix is "hollow" – and inversely, if the pieces are large but few in number, all interconnected, the matrix will be small with crosses everywhere. This lets us visualize the issue of granularity, whose importance we have also recalled.

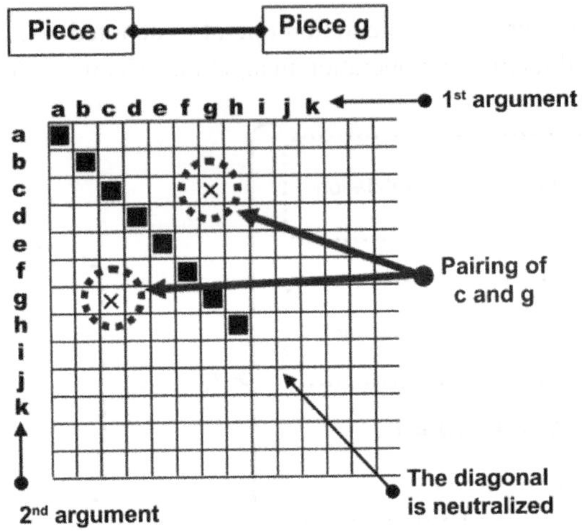

Figure 8.13. *Coupling matrix*

The matrix lets us choose pieces that have neighboring relationships with each other such that natural groups or series are created – in systems jargon, they are clusters/building blocks – in order to avoid creating an unreasonable amount of work to reconstitute a validation environment that is close to reality.

Systemic analysis makes it possible to develop an integration strategy, that is, a method for grouping elementary pieces which will give the value of the number n. If n is large, all N pieces could be gathered in one block, and then everything would immediately fall into the chaos of the combinatorial approach developed in Part 2. There are too many errors to consider simultaneously. If n is small, maybe 1 or 2, a large number of validation environments will need to be created and will by definition be redundant, therefore with a low yield. With $n=1$, as many environments are needed as there are pieces, or N, and with $n=2$, the number of first-level environments is exactly $\frac{N}{2}$, whence $\log_2(N)$ layers of groups, as in the tree from Figure 8.3.

Making environments also means introducing new errors, which means that in the case where a new error is discovered, we must make certain that the error is due to the piece itself and not the environment created to validate it, or a conjunction of the two. Systemic analysis will help us find a path in the pairing matrix, starting with an initial granularity and building step by step on already validated points so as to reuse environments until the final, integrated system is reached and the pairing matrix is reduced to one block. The issue of the computer stack presented in Part 1 is probably the best example, in any case the most accessible one, to illustrate this systemic issue of quality without which it would have been unimaginable to construct interfaced architectures from silicon to the point of the W3C cyberspace protocols that are present in all our smartphones.

We have a fairly precise idea of the evolution of the cost of errors over time; see Figure 8.14, which contains a qualitative curve that does not account for scale. At first, the introduction of new technology such as the use of the digital to replace our dependence on analog had a cost, that of innovation. This cost would necessarily decrease as people learned about it and it matured, until it reached an irreducible cost.

The PESTEL environment means that the curve will be subject to hazards that indicate how the organization adapts to innovations; with IT, it was a succession of variously intense shocks, such as the introduction of workstations, personal computers, client/server architectures, Internet, etc., over the course of the 40 years of Moore's law. Each time, the cost rises again until it eventually gets out of control, following errors by a manager who does not understand these repeated innovations or who gives into the marketing strategies of technological vendors who promise miracles. The jolts are at times destructive and affect the company culture, as with the arrival of the Internet. The end of the life of a system is generally characterized by an unstoppable increase in the cost of errors that can no longer be compensated for; the system has reached its limits of its evolutionary and adaptive capacities. Now comes the service retirement and organized dismantling (see Figure 8.9), whose rules are also part of the reference.

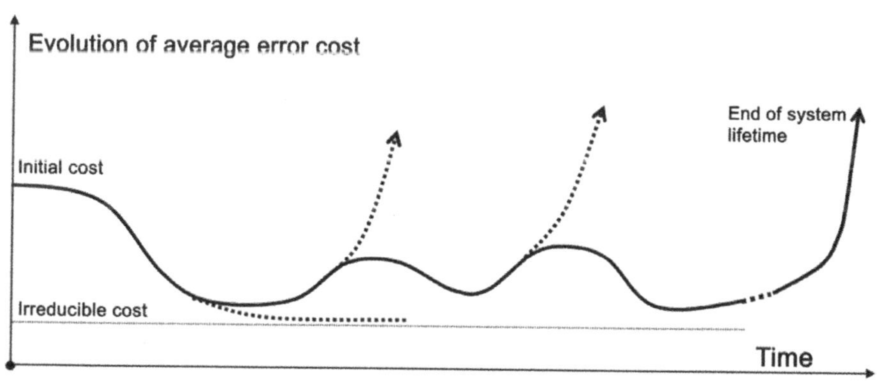

Figure 8.14. *Evolution of error costs*

Everything that determines and influences the cost must be identified and modeled; for that, systemic analysis and its focus on retroactive cycles are irreplaceable. A recent example is that of the tests which some people thought they could delegate to sub-contractors in low-cost countries such as India. This had the almost immediate consequence of deflecting responsibility from developers, who are much less vigilant towards their own errors, whose rate then rises. Moreover, the validation environment gets lost, so the capitalization of investments and the increase in maturity are lost. In the end, the complete cost rises, and the quality crashes.

8.3.3. *The systemic lesson of projects*

A systemic approach to the science of projects shows, and this is incontrovertible, that the complexity of usage and the complexity of engineering in their human dimension must be addressed seriously; these are two domains where incertitude and ignorance, in the sense of being confronted with phenomena that we cannot predict or even imagine, incontestably reign. Whether it is running out of rare resources, the ecological problem of managing our trash which accumulates beyond what is reasonable, or in a very different domain, the consequences of the demographic shock that Africa will experience in the next 30 years, an open world, preferable to any other, is an uncertain world where we must advance carefully.

As the science of projects shows, the only weapons at our disposal are those of retroaction and control, since going back in time to how it was before is impossible, as we can see in physics and its "simple" systems from which mankind is excluded – all the more so in human systems where survival means adaptation and learning. To retroact with intelligence, we must predict what the future will be, staying one or two steps ahead, which will allow us to judge both the speed and the acceleration of phenomena.

Jacques Lesourne, a professor at the CNAM in the 1990s, in an already outdated book where he discusses systemics applied to organizations and forecasting, *Les systèmes du destin* (1976, so before the digital revolution of the 1990s, but it was already known at this time that the impact of TIC was going to be enormous), underscored a major point whose meaning is ever more apparent today: the "insufficiency of controls" in the last two chapters of the book, Chapter 10 "A Model of Contemporary History" and Chapter 11 "A Synthesis for Hope". The more complex it is, the more controls are needed, but the control he speaks of is not the police control whose ravages have been experienced by all – that of von Neuman's "who watches the watchmen?" – but that of systemics, which we clearly see at work in the implementation of project science. As with war, the science of projects is an art of execution, to paraphrase Napoleon.

The main lesson of errors is that we know that they exist: it is an absolute certainty, but we do not know, by definition, neither where they are – unavoidable ignorance – nor what their consequences will be – the uncertainty of facing the hazards of an environment – depending on the contexts of use. In the situations that complexity automatically creates for us, the project is at its ontological endpoint, fundamentally, a "machine" for detecting errors to escape the insignificance of the combinatorial, since before even imagining a correction, that is, a retroaction, we must know exactly where we are in order to act appropriately. This is what the third law of the *Compendium* stipulates, and without it, without its implementation "at the same time", the first two will not be operative. To do something, even when the desired objective is precise and rigorously defined, is also to feel around, to proceed by trial and error in order to have the control cycle emerge, a cycle that is hidden in the details, to identify the right compromises that will lead to specificities that will enable regulation.

To conclude, we can say that systemics, as illustrated in this book, shows that it is a fundamental point of reference for organizing the complexity that a project makes visible.

Conclusion

Prospective Dialogue

This conclusion comes following an interview between Philippe Kourilsky, Emeritus Professor of the Collège de France, and Jacques Printz, concerning an intersectional perspective on his book, *Le jeu du hasard et de la complexité* (or *The Game of Chance and Complexity* in English, published by Odile Jacob in 2014), and the fundamental questions raised by the book in terms of the complexity of living organisms, with a focus on the immune system and the way the complexity of artificial systems designed by engineers are understood through a "systemic" lens, as presented in this book.

As we have seen, complexity has entered the domain of the sciences by being mathematized through the foundational work by von Neumann, work that in the 1970s–1980s ended up leading to an entire corpus that is now a huge part of the theoretical computer science known as complexity theory. This corpus is inseparable, although quite distinct, from the problems posed by information engineering, with which it is in symbiosis. Today, information is at the heart of systems, along with complexity, from which it is inseparable.

Parallel to this, but without any direct relationship, the study of genetic code gave rise to a reflection on the complexity of life, with works like those by von Förster and notions like "order by noise", which sound strange to engineers, where the opposite is what is affirmed, as in Shannon's 2nd theorem on the channel with noise that led to the error correction codes that compensate for that very noise.

One of the flagship books from this period, after that by Monod, *Chance and Necessity* (1972), was the book by Atlan, *L'organisation biologique et la théorie de*

For a color version of all the figures in this chapter, see www.iste.co.uk/printz/pedagogy.zip.

l'information (1972), followed shortly thereafter by *Entre le cristal et la fumée* (1979). Based on these, the dialogue between bioengineering and systems engineering had become difficult, if not impossible. But very quickly, many biologists realized that these slightly overly qualitative notions were not productive; they did not provide a way of understanding the complexity of living organisms. Atlan himself wrote about the state of the field in a small book, *La fin du tout génétique? Vers de nouveaux paradigmes en biologie* (1999).

With the book by Kourilsky, this page seems to have fully turned. The dialogue should be able to restart on a foundation that integrates complexity into the heart of discussions. In the 1970s–1980s, the scale of complexity for engineers was, in terms of the number of elements, let us say in the order of hundreds of thousands, on the margin of a million.

It is now, let us say since the 2000s, in the range of a billion; the chip from the NVIDIA QADRO GV100 computer contains 21 billion. The electronic components are of nanometric size, completely invisible to the naked eye, and their interconnections so delicate that once built, they can no longer be taken apart even if we are simultaneously able to demonstrate that they do function and diagnose errors that might occur with the help of ad hoc mechanisms that are completely controlled via a so-called PAC logic, whose origin can be found in the work of von Neumann.

We will conclude this book by exploring the resumption of this dialogue.

Engineers in the face of the complexity of living things

In this book, we have voluntarily refrained from making any references to issues in biology in order to keep to a strictly productive approach, which is that of engineers.

The physicist Richard Feynman used to say he did not "understand what he did not know how to recreate and/or build"[1]. The temptation to use metaphors and/or analogies from biology to illustrate certain aspects of systems that engineers create is constant, but this is not a good approach. A metaphor is not efficient unless it illuminates an obscure point. But with biology, we are in the contrary situation. Do we understand how living organisms work, beginning with cells, in the way we understand the most complex artifacts invented by human ingenuity? Philippe Kourilsky points out that understanding a living organism such as the human means

1 A formula found on his blackboard at the time of his death, in February 1988, taken from a photo in the archives at Caltech.

understanding the interactions of approximately 10^{13} micromachines that make up the cells of a human adult; this means understanding ontogenesis, that is, what makes one unique cell go from the fusion of one ovum and one spermatozoid to a complete human who will only be an adult and able to reproduce after about 12–13 years, plus some more years before being able to master the cultural codes of increasingly complex societies. It also means understanding their renewal, at variable rhythms depending on the organs, and therefore their elimination, throughout a lifetime, which for humans does not seem to be able to extend beyond 120 years. Finally, it means understanding the interactions between the bacterial flora we carry with us, a flora that is as numerous as our own cells, and with which we live in symbiosis.

The task is immense! The abilities of living creatures fascinated scientists such as von Neumann, Shannon, Turing and Wiener – those who were present at the start of systems sciences, that is, cybernetics, with its flagship machine, the computer.

It is now predicted that by around 2025, there will be 50 billion objects connected to the Internet, or "only" 5×10^{10} devices, which are certainly much smaller than the ones they knew during their lifetimes, but which have nothing to do with the size and complexity of cells, which are miniature chemical factories.

The engineering of engineers is therefore still very far from the one that is at the heart of the living. Nevertheless, it seems to us that it is beginning to have some points of contact.

When Philippe declares that to live, we must first survive, this is something that speaks to a systemic engineer, worried about ensuring that the complex technical objects he will be passing on to putative users will keep working.

And when Philippe underscores the importance of modularity for understanding living things, the echo to engineering is immediate since the fundamental work of a systemic engineer is to break down an object he wishes to make into a set of individualized modules whose integration, that is, the process of construction (in the language of biology, this is called ontogenesis), will answer the needs formulated by users with a guaranteed reliability in a given usage environment that can vary.

Behind this question of modularity, the entire issue of architecture becomes apparent; it is the very core of systems engineering.

It is on the basis of these points of convergence that now seem within our reach that we envisaged this interview/discussion.

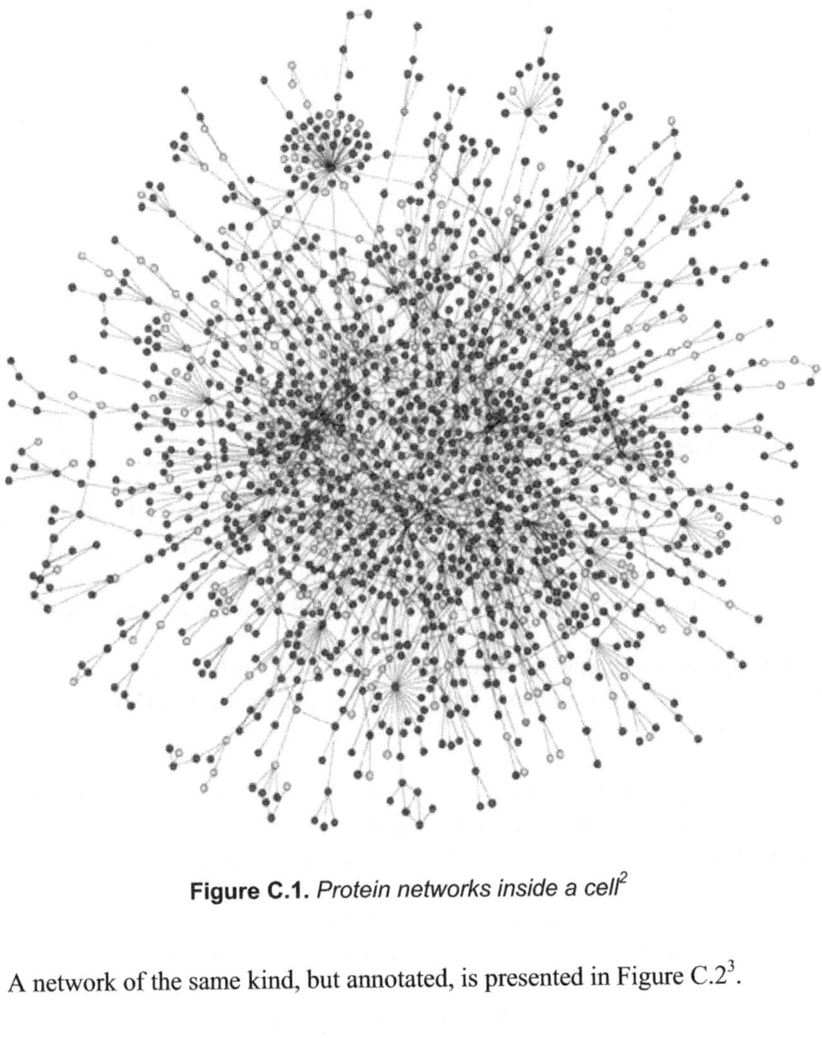

Figure C.1. *Protein networks inside a cell*[2]

A network of the same kind, but annotated, is presented in Figure C.2[3].

2 Source: Barabasi, A.-L. et al. (2004). Network biology: Understanding the cell's functional organization. *Nat. Rev. Genet.*, 5(2), 101–113.

3 Source: Guruharsha, K.G. et al. (2011). A protein complex network of Drosophila melanogaster. *Cell*, 147(3), 690–703. The drawings in Figures C.1 and C.2 can be downloaded from: http://www.cs.cmu.edu/%7Eepxing/Class/10810-06/readings/Barabasi_Oltvai_04.pdf and https://web.stanford.edu/class/gene211/pdfs/Guruharsha-Drosophila-proteincomplex-network.pdf.

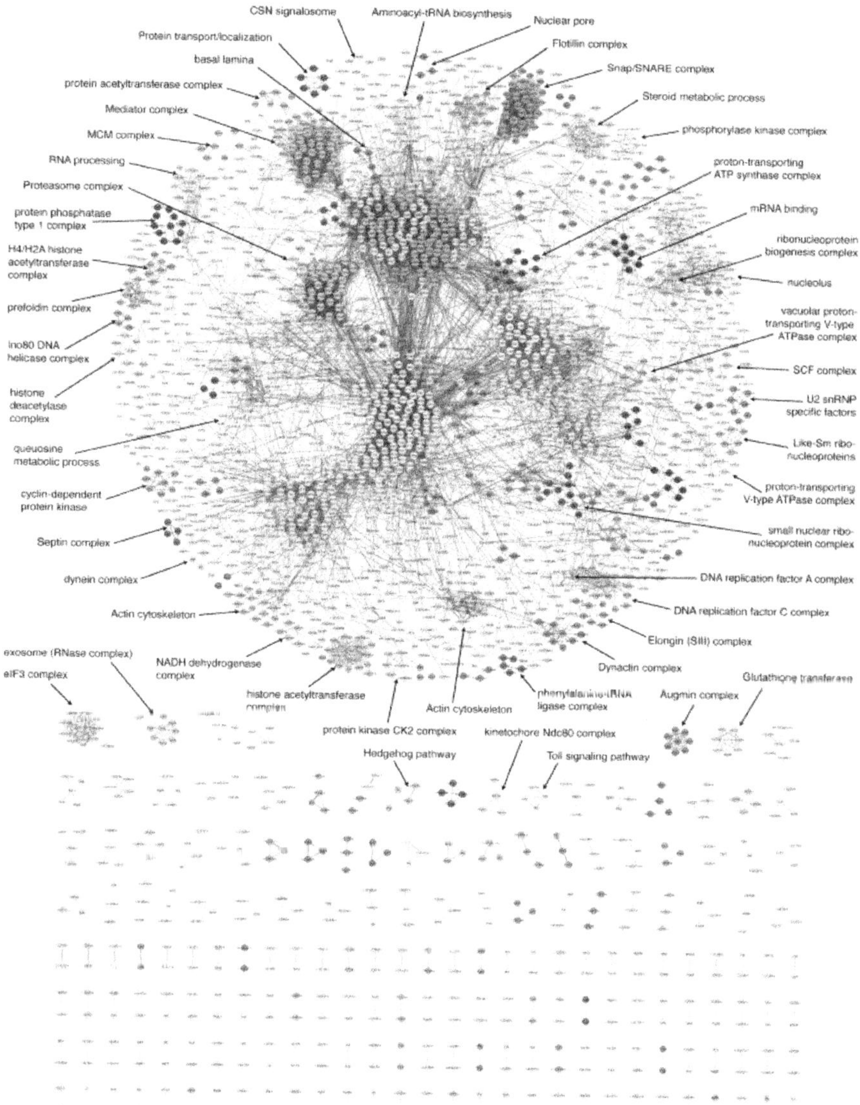

Figure C.2. *Modules, annotation and modeling*

engineering science, namely, ensuring proper functioning, which extends, in the area of information, to one of the "two new sciences" identified by Galileo: the resistance of materials, in other words, the why and how of "it breaks". It seems to me that in this, there is an entire area of knowledge that is perhaps common to the science of engineering and bioengineering, if we make some effort towards abstraction. Von Neumann, the principal inventor of modern computers, was fascinated by the reliability of an organ like the human brain; he even based his final book on it, *The Computer and the Brain*, just before his death, in 1957. How do you see all of this?

Response by Philippe Kourilsky

Indeed, we must first survive the hazards of life, especially microbes; recall the children born with a serious immuno-deficiency and who needed to be placed in a bubble to protect them from the environment. We live in a sea of microbes which we do not see but that are everywhere: on our mucous membranes, in our lungs, in our intestines, which are the primary reservoir for them. There are approximately as many bacteria in a human organism as there are cells. They are about 100 times smaller, but that still makes up about a kilo. Some of these microbes are pathogens. Pasteur was the one who first understood their role in infections. The organism must defend itself; otherwise, the infectious agents, parasites, bacteria and viruses, will ravage the system. These are human's greatest predators. These infections are essentially unforeseeable, but the probability of being infected depends on the ecosystem in which the organism is immersed and where it pulls its energy from. On the contrary, if the organism is infected once, it will establish defenses that, next time, will be faster and more efficient.

But errors are a second manifestation of randomness. Let us take the example of coagulation: we can fall and get hurt, often by chance; the blood flows and the organism initiates the coagulation mechanism, which will progressively close the hole made. But if, by an error that also arises from chance, this mechanism is wrongly activated, and a coagulation makes a clot in the heart or the brain, we will die!

The nature of chance that is closest to what you do in engineering is that of internal function errors. We have seen them at some levels but we can hypothesize that they are present at all levels. The replication mechanism for DNA is activated at each cellular division; there will be replication errors whose frequency we could estimate to be 10^{-9}, but, given the number of cells that replicate, in total, that ends up being a lot. In fact, it is improbable that the DNA of two cells of any one organism is entirely identical. That does not immediately have consequences for survival. But over time, the accumulation of errors ends up causing, at least partially, aging, and, too often, cancers. We should also note that over the long term, these "errors" are also the motor for the evolution of the species.

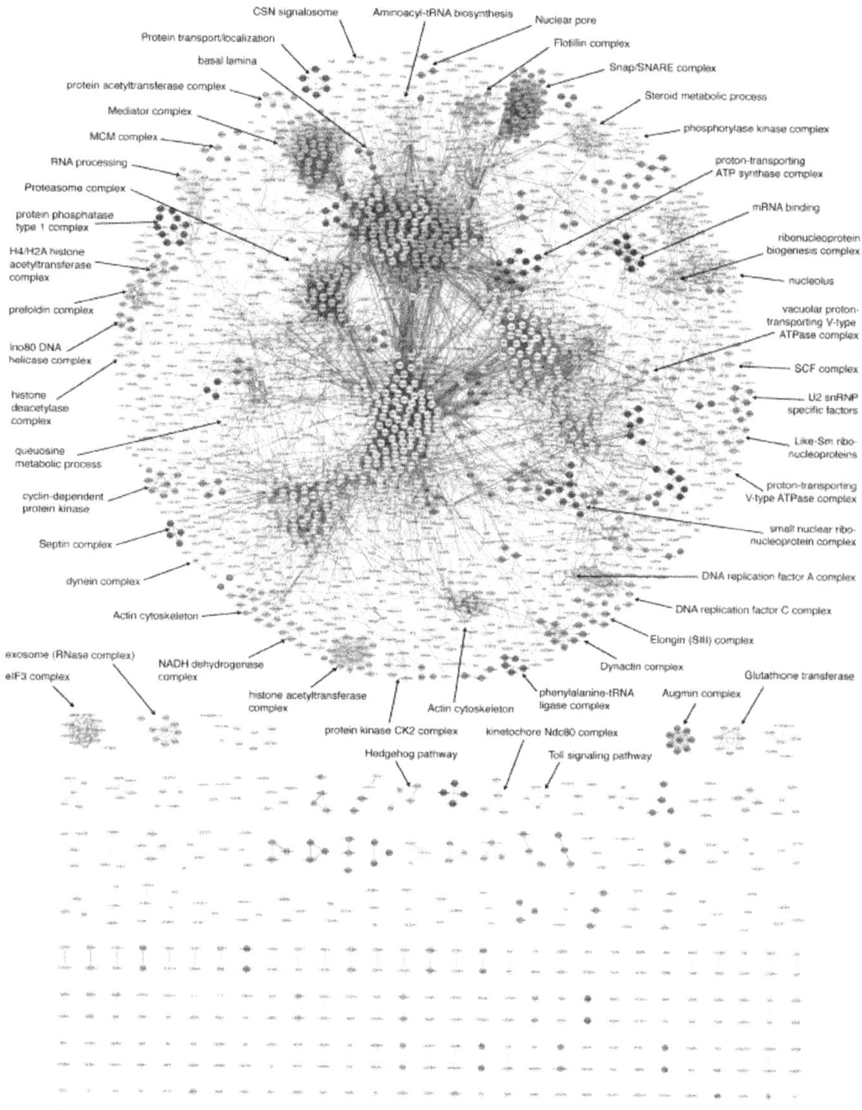

Figure C.2. *Modules, annotation and modeling*

Can we compare the engineering of life and that of engineers? Six questions for Philippe Kourilsky

Question no. 1: engineering

In your last book (Kourilsky 2014), you refer to the engineering of engineers, in particular with reference to articles by John Doyle. Engineering sciences have constituted, for two centuries now, a corpus of knowledge that made the Industrial Revolution possible, carried by the "Grandes Ecoles" in France, especially "X" (Polytechnique), Centrale (Engineering school of the University Paris-Saclay), etc. Is there something comparable in biology? For example, in the area that we call bioinformatics? For engineers, everything they do is finalized with what we call projects, for which we could say systemics, in the sense we give it in this book, is the language.

Response by Philippe Kourilsky

In the recent history of biology, there have been several stages and important turning points. An initial turning point was that of biochemistry, propelled notably by Louis Pasteur, who was trained as a chemist. He introduced the rigorous, scientific approach to the study of living things by pushing back against vitalism. A second turning point, just as important, was the discovery of DNA and its structure. This happened in the 1950s, thanks to analytic methods using X-rays, imported this time from the physics of high energy. These discoveries made it possible to grasp the double-helix structure of DNA and understand how genes and the chromosomes that carry them are replicated. Shortly after, it was the genetic code that got elucidated. With the four elementary building blocks of DNA, the four bases ATCG, 64 combinations of three letters can be made, 64 codons associated with the 21 amino acids, which are themselves the building blocks of proteins. We can thus decipher the dictionary of genes → proteins. Hence the emergence of molecular genetics, in the 1970s–1980s, then, with the rise in power of instruments making it possible to sequence DNA, that of bioinformatics. In the interval, computers became powerful enough to, on the one hand, save entire genome sequences, and, on the other, to analyze them. For a human genome, this means approximately 3.4 billion base pairs, or, in the norms of paper editing, the equivalent of 850 volumes of 1,000 pages each. The analysis of genomes is therefore completely impossible to the naked eye, but with computers, whose performance hit a boom in the 1990s, what seemed inaccessible became possible, and biologists along with the help of programmers were able to do research on patterns, sorting, comparisons, etc., in genomes. In short, they were able to make use of the information collected thanks to instruments that were themselves computerized.

The work of a biologist, until the arrival of computers, was a job at the "lab bench" where the majority of time spent consisted of preparing and following experiments that would take months, or even years, with standardized experimental protocols to guarantee the reproducibility of experiments and the statistical processing of results.

The arrival of computing has changed much of that today, so for many researchers in biology, the lab bench no longer makes up more than 30–50% of their activities, with the rest of their time being reserved for analysis and researching information, most often via computational means. The devices and equipment that biologists use today for sequencing DNA, obtaining images of cells, etc., have the characteristic of producing large quantities of data, *Big Data*, which is impossible to understand without the assistance of models. To begin with, they must be able to be stored and indexed so as to easily handle them in the equivalent of libraries, in the common sense of the term.

When we say that two human genomes are identical to 1/1,000 (which is correct), this implies that 3.4 million bases are distinct from one individual to the next. These differences are distributed in enormous combinatorials. Therefore, all genomes, even within the same species, are different. It is in analyzing all of this information and deciphering the impact of these combinatorials that programmers-turned-bioinformaticians have found their place in teams of biologists. For example, in an institute that I created and presided over in Singapore, very advanced in terms of biotechnology, there were a dozen or so highly qualified bioinformaticians for 200 researchers. What gets called *systemic biology* or *systems biology* includes a large proportion of bioinformatics, and in this sense, we cannot compare it to systemics such as is presented in your book. Nevertheless, the articles by John Doyle that I sent you were a bit of a revelation for me[4]. I concluded that it was necessary for biologists to take an interest in the engineering of engineers.

Question no. 2: survival

In your book, there is a formula that especially caught my attention: "To live, you must survive". You develop notions of robustness (which others call resilience) around this idea; you speak a fair amount about errors, and you present the immune system as part of a vast, general defense system against all hazards, internal and external, that living organisms are subjected to. You also speak of "chance", which you qualify, in order to make the distinction, as saturable chance and non-saturable chance. All of this evokes, for the engineer in me, the most fundamental areas in

4 This refers to "Reverse Engineering of Biological Complexity" and "Architecture, constraints, and behavior," both downloadable.

engineering science, namely, ensuring proper functioning, which extends, in the area of information, to one of the “two new sciences” identified by Galileo: the resistance of materials, in other words, the why and how of “it breaks”. It seems to me that in this, there is an entire area of knowledge that is perhaps common to the science of engineering and bioengineering, if we make some effort towards abstraction. Von Neumann, the principal inventor of modern computers, was fascinated by the reliability of an organ like the human brain; he even based his final book on it, *The Computer and the Brain*, just before his death, in 1957. How do you see all of this?

Response by Philippe Kourilsky

Indeed, we must first survive the hazards of life, especially microbes; recall the children born with a serious immuno-deficiency and who needed to be placed in a bubble to protect them from the environment. We live in a sea of microbes which we do not see but that are everywhere: on our mucous membranes, in our lungs, in our intestines, which are the primary reservoir for them. There are approximately as many bacteria in a human organism as there are cells. They are about 100 times smaller, but that still makes up about a kilo. Some of these microbes are pathogens. Pasteur was the one who first understood their role in infections. The organism must defend itself; otherwise, the infectious agents, parasites, bacteria and viruses, will ravage the system. These are human’s greatest predators. These infections are essentially unforeseeable, but the probability of being infected depends on the ecosystem in which the organism is immersed and where it pulls its energy from. On the contrary, if the organism is infected once, it will establish defenses that, next time, will be faster and more efficient.

But errors are a second manifestation of randomness. Let us take the example of coagulation: we can fall and get hurt, often by chance; the blood flows and the organism initiates the coagulation mechanism, which will progressively close the hole made. But if, by an error that also arises from chance, this mechanism is wrongly activated, and a coagulation makes a clot in the heart or the brain, we will die!

The nature of chance that is closest to what you do in engineering is that of internal function errors. We have seen them at some levels but we can hypothesize that they are present at all levels. The replication mechanism for DNA is activated at each cellular division; there will be replication errors whose frequency we could estimate to be 10^{-9}, but, given the number of cells that replicate, in total, that ends up being a lot. In fact, it is improbable that the DNA of two cells of any one organism is entirely identical. That does not immediately have consequences for survival. But over time, the accumulation of errors ends up causing, at least partially, aging, and, too often, cancers. We should also note that over the long term, these “errors” are also the motor for the evolution of the species.

When a gene "expresses" itself, that is, it leads to the synthesis of a protein whose code it carries, there can be errors in the transcription of the gene to the RNA, then in the translation of this RNA into proteins, which are made up of amino acids. An average protein is made through the linking of hundreds of amino acids. We estimate that the rate of errors in this chain is one error for every hundred thousand incorporated amino acids, or about 10^{-5}. Over hundreds of thousands of proteins, this ends up being numerous and is significant.

All cells communicate with each other. In particular, many cells circulate and exchange chemical messages with each other. This implies that there are systems for addressing each other and/or recognition systems that mean that the protein (or any other molecule) emitted as a signal will, or will not, arrive at the correct destination, or perhaps that the mechanism for associating it with the correct receptor will, or will not, be produced correctly. Thus, the "right" receptor can receive an incongruous molecule. There are therefore thousands of opportunities for this not to work or to work incorrectly.

The way in which the organism organizes itself to survive is that, at every functional level, we find error correction mechanisms. At the level of DNA synthesis, the primary copying mechanism is precise to 10^{-3}, and it is associated with two types of correction mechanisms. Each one has a reliability of approximately 10^{-3}, which means that in total, that makes for 10^{-9}. Likewise, at the level of protein synthesis, there are a series of measures that expel a bad amino acid at the moment of incorporation, if it is not the right one. But since the number of steps is the highest for DNA, the probability that errors will occur is, in the end, higher (10^{-5} instead of 10^{-9}).

I would emphasize the fact that even if errors are so infrequent that they are not observable throughout the life of an individual, we must not think that there are not any. On the contrary, there are many, all the time, at every moment, but the immense majority of them are corrected. Why are 100% of them not eliminated? Here, a very important point is introduced, which is the economy of the system. Eliminating more than "necessary" would end up costing too much in terms of energy and time for repair, which would mean that the organism would no longer be competitive and would end up being eliminated. In fact, the rate of errors observed are compromises that are carefully selected by evolution, which does not mean they are unique, as there are specific compromises in each individual and by design, the individual must be competitive enough to survive. The economy of the biological system means that the system cannot be improved beyond a certain threshold. The temporal aspect is also important, since the time of response to correct errors and the energy needed are obviously elements of survival for the organism.

What I call saturable/non-saturable chance is a question in itself! What we observe, is that, in our blood, at every instant, there is something in the range of 10^9 different antibodies, made randomly by our immune system. But the space of possibilities is much vaster: it is in the range of 10^{18} since it comes from the random recombination of several series of genes. It is the same for another class of receptors that are found on the surface of other immune cells, T-cells. Their real repertoire has approximately the same diversity (10^9), but it is moreover modeled by molecules from the major histocompatibility complex (in humans this is the HLA system). These are also very diverse and make for a combinatorial such that it is difficult to find two "histocompatible" individuals for organ transplants. In total, from an immunity standpoint, we are all different! Nevertheless, our immune systems function in the same way overall, with fairly comparable results (we manage to vaccinate 90 or 95% of a population, but there is a small number of individuals for whom the vaccine does not work).

Experience shows that we can get an animal to produce specific antibodies against almost any molecule. Generally speaking, we can argue that the system manages to synthesize antibodies against anything, even against molecules that have not yet been imagined by chemists. Biologists say, in their jargon, that the ensemble of antibody structures saturates the space of forms that they are able to recognize. The underlying mechanism makes use of at least two random measures. The first is based on the random recombination of gene segments. The second relies on the capacity of a B cell that produces antibodies of modest quality (affinity) to trigger, when exposed to the antigen a second time, an explosion of mutations of which some might improve the capacity for recognition, and they will be selected because of this. These mutation/selection mechanisms improve the response in a remarkable way.

A third type of mechanism plays an essential role. It is the capacity of immune system cells to reproduce rapidly. In the engineering of the living, components can multiply, which is not the case for the engineering of engineers. A vaccine will induce an accelerated multiplication of cells that carry the useful antibody. In the end, we can end up with billions of cells to combat a viral or other kind of invasion. There is a race between the development of the organism's response and the infectious agents that are themselves capable of self-reproduction. At the start, the so-called adaptive system is not efficient enough (it is supplemented by "innate" immune responses), but after a certain number of days, it becomes capable of controlling the infectious situation.

This brings me to a very important point. I explained that randomly produced antibodies saturate the space of forms, and this is what I call saturable chance. But, for this effect of saturation to occur, there is a hidden condition: the degeneration of specificity. In biology we attribute the qualification of the "specific" to a process that brings about a given result in a given amount of time with an error rate that is considered weak, and even indetectable, but it is entirely relative. In other words, we would qualify antibodies as very specific if they recognize one molecule among thousands of others. But billions rather than hundreds of billions, millions or thousands? And what happens in reality? Confronted with a soup of molecules, an antibody can easily recognize one with very high affinity, and several others with less affinity. Its specificity is thus a function of the way we measure recognition. Or even, an antibody recognizes two molecules with a strong affinity but the existence of one of them escaped the experimenter who perhaps thought that the antibody was absolutely specific.

Let us return to our billion antibodies that saturate the space of recognizable forms. The dimensions themselves of this space depend on the granulometry that is adopted, but we can consider it far superior to the number of antibodies, which in humans is a few billion. This implies that the antibodies saturate the recognition of this space due to all sorts of coinciding reactions. Incidentally, the antibodies of a mouse, less numerous and less diverse than those found in humans, also saturate the space of forms. The definition of specificity therefore depends on what we consider to be specific; that can seem circular, but that is how it is. The same antibody is capable of recognizing several different structures, but we do not know how many; it depends on the window of interactions we use to observe, but all of this contributes to the saturation of the space of forms. What defines it is that it works!

This question of specificity can be posed – in the same manner and with as much importance – about the networks of reactions that operate in every cell (I had not thought of it from this perspective until our discussion). In each cell, there are an enormous number of biological chain reactions at work. There are almost certainly opportunities for coinciding reactions, therefore potential sources of errors, but the ensemble is adjusted (selected) so that "it works". It is a chain of elementary reactions that at a certain moment will provoke an effect/result in the cell, often depending on its environment. A cell is like a telecommunications satellite with its antennas standing in all directions, with its receivers and emitters. Receivers will at some point capture a molecule emitted by another cell with which an interaction is possible. How will what happens at the surface trigger the expression of a gene? It is with the help of a chemical cascade called a signal transduction that, from the initial interaction, will, through a series of steps (typically a dozen or so), propagate captured information towards the inside of the cell. At each stage, specific reactions are at play, and there are possible errors. We must therefore ask ourselves if there are mechanisms for error correction at each stage. Very often, we do not know enough to detect them, especially since there can be very subtle changes in the spatial anatomy of proteins. In the end, there is a certain probability for error in the transmission of a signal. This is similar to what happens in the synthesis of proteins, which I have already spoken about, where the error rate is higher than 10^{-5}.

There are many other possible errors, especially the ones that affect the spatial organization of proteins. We estimate that up to 30% of proteins that have just been synthesized are potentially ill-formed, which is huge. Thus, in vivo, we find quality controls that monitor and facilitate the assemblage. All of that creates waste, which is either reformatted, or eliminated by a complex pathway system that identifies what is defective and destroys/recycles what cannot be repaired. Organs like the spleen and the kidneys thus play a major role in the organism. There is in this it seems a significant difference with classical engineering systems that do not use such a sophisticated pathway, present at many levels of the system.

As I understand it, the study of this mechanism does not occupy enough space in the way biology is taught; we are happy to acknowledge that we have organs for taking care of this filtration: kidneys, spleen, etc., and we do not ask enough theoretical and practical questions. We must also keep in mind that our organism constantly regenerates a large part of its components, even if at different rhythms: our body is almost entirely renewed, let us say every 10–20 years, which does not imply the death and the synthesis of hundreds of billions of cells per day. The proper functioning of the mechanisms that manage our internal waste disposal is capital; if they fail or work improperly, the organism can die from poisoning.

Question no. 3: modularity

You speak a lot about modularity, which is another fundamental notion in artificial systems engineering, and to which the notions of hierarchical constructions and interfaces are directly connected. In the engineering of engineers, the hierarchy of the elements in a system and the rules for assemblage, what we call integration in our jargon, are truly at the foundation of our activity, since without that none of the systems that are so familiar to us today, such as smartphones, could exist. This has even been theorized by physicists like Anderson in his short article from 1972, "More is different"[5], where each level of integration has its own but overall coherent logic. This article was the starting point for what physicists called the "physics of emergence" (a very good introductory book is the one by the 1998 Nobel Laureate Laughlin, *A Different Universe*[6]). This is also at the heart of our book. Shouldn't there also be, here too, points of contact between life sciences and information sciences? This notion of module is to be used with great precaution since, in systems engineering, as in biology, it seems to me that there are modules in the functional sense of the term but there are also properties that cannot be captured by this notion; for example, robustness, or performance. In no artificial system do we find a "box" that could be called robustness: it is a property that is spread across all the functional

5 *Science*, 177(4047), 393–396, 1972.

6 Published by Basic Books in 2006.

modules that depends on the nature of the interactions. Information concerning robustness pervades the entire "environment" of the system in the manner of a field, in quantum physics, with its very particular particles such as the very high-energy but very fleeting bosons. Engineers thus clearly distinguish what they call functional characteristics from so-called non-functional characteristics, linked to general properties. In our field, there is an article by David Parnas who became famous for this notion of modularity in 1972 with "On the criteria to be used in decomposing systems into modules"[7]. Where are we at in life sciences? Survival is an overall property that is distributed all over, that consumes energy, that interacts, but the real problem is the *how* of the corresponding engineering? What can you say about this subject?

Response by Philippe Kourilsky

The problem of modularity in biology is, conceptually, even philosophically, complicated. The profound reason is that modules are not obvious. A biologist is faced with very complex organisms; their problem is how to determine subsets that make sense. On the anatomical end, it is fairly clear: macroscopic organs like the liver, a muscle or a bone constitute modules connected by the major communication pathways of the nervous system, the circulation system and lymphatic circuits. We are in a classical architecture. When we dissect an organism and we see what happens, we can use morphological criteria as a starting point, but when we reach the level of cells, it becomes much more complicated since the entanglement of constituents is enormous.

How can we appreciate the complexity of a system? We can, of course, list the elements in it and the different links regulating their interactions. But we also tried to evaluate it using control theory. We thus wonder about the number and the proportion of nodes in a network that must be under control to make the system go from state A to state B in a finite amount of time. The exercise was led over several dozen networks described in scientific journals. The result, which I find interesting, is that, in dealing with biological networks (such as the one in Figure C.1, for only one cell), 70–80% of the network nodes must be controlled, something that certain experiments tend to corroborate. In comparison, for an electricity distribution network such as the one in Europe, it seems that 30% is sufficient. This difference suggests that the architecture of biological networks is unique.

In terms of modularity, we must thus ask ourselves how we define the modules in a system whose elements are interconnected at 70–80%. What researchers do is annotate the network of interdependencies (see Figure C.2, which concentrates years of biological knowledge). For example, when we see that 10 proteins have participated in the replication of DNA, we are tempted to make a module of it

7 *Communications of the ACM*, 15, 1972.

projects that designed and developed them reached a limit beyond which, if we overstep it, failure is certain, since we do not know how to control the combinatorial complexity induced by the size (approximately 1,000 engineers). The "infeasible" system, in the sense of engineering, decomposes into subsystems whose integration constitutes the initially imagined system, and this happens over a succession of levels that interlock with each other, each of the levels preserving its capacity to be engineered. This is what we call interoperability, or a system of systems that cooperate via an exchange model. It seems obvious to us that this notion is present everywhere in bioengineering, starting with the cell, most likely, which constantly interacts with its environment. How did you understand this part of our book? And how could we imagine a use for biologists, even at the cost of changing the language/vocabulary so they can accept it?

Response by Philippe Kourilsky

In biology, it is fairly simple since the human organism, for example, can be described by a set of 10^{13} computers that interact. Each cell is a computer and what unites them is that they all share the same DNA, notwithstanding the previously described errors. From time to time, one of these cellular computers deviates slightly, and sometimes totally. The question is thus to understand how the organizing principles of these interactions are described in the DNA, but, for the moment, we do not know much about it!

The environment intervenes, not only in the form of the more or less dangerous aggressions that we spoke of earlier, but also to model the system, depending on the interactions it encounters. It is especially notable in the case of the brain, which is very sensitive to interactions and the history of the individual, as well as for the immune system, which integrates the experience of its construction and the hazards it encounters. In another area, nutrition (and/or malnutrition) influences the construction of muscles and bones. In the biological human being, even if we disregard the brain and consciousness, which correspond to a different logic, we obviously find ourselves with the 10^{13} computers operating together, in a configuration of systems of systems, with all of the functional hierarchy.

Another important point to highlight is that this system is not only communicative; it does things, it acts, internally and externally. It resembles an automaton more than a computer. This system carries out eliminations as I explained earlier; it kills, it renews, it synthesizes, etc.

We can also take the question to a different scale, from the perspective of the evolution of species and the interoperability that this carries with it. Over the course of the modifications of reproducing organisms, some end up no longer being able to reproduce; we thus find ourselves with two species that are thus no longer interoperable, that is, interfertile (which is a definition of a species). We must also take structural and functional redundancies into account as they preserve a certain measure of interoperability: for example, if a gene drifts and two copies of it exist, the system will remain generally operable and interoperable. We should note that redundancy is an element of robustness, which prevents or limits the expression of errors, continuing until the ontogenesis of the organism.

Concerning what you call exchange models, this also refers to ontogenesis. In the living, at the first level, we can say that everything is conceived of to be interoperable. When that begins to drift, as in the case of a cancerous cell, there are mechanisms that control the proper functioning of all that, and they will limit the deviant cell. This is, for example, the case with lymphocytes called *natural killers*, which are constantly armed to kill, but are kept from doing so by characteristics of normality. As soon as these disappear from a cell, it is destroyed. When these mechanisms fail, for *x* reason, that is when the cancer develops. When two organisms diverge, they lose the property of being interoperable while continuing to share the same ecosystem. We can therefore look at what happens in the ecosystem taken as a whole, and there we will find both aggression, via every kind of infection agent and everywhere, and cooperation, also everywhere, beginning with what happens in our intestines, where billions of bacteria take part in the digestive and metabolic processes; this is the same for the roots of plants.

Question no. 5: quantification

You also speak of the "biology of quanta", and very carefully moreover, but here again, it is something very fundamental, especially in information science. The computer is fundamentally a machine with states, and that is the reason why von Neumann created a new science, so he could understand the theory of automata (in 1948, to be precise), of which your colleagues from the Académie, like Maurice Nivat or Joseph Sifakis, are the most eminent representatives, in France. The entire art of systems engineering is to find the right level of magnification for "information quanta", which are the ones we have access to via the interfaces of our smartphones, tablets or command screens, for behind the interface, there is always a computer. It is the metaphor for the computer stack, which is also a central element of our book. What does that evoke for you, especially in the immune system, which activates depending on what it encounters, again, in a way that reveals several similarities with what happens in an artificial system? Behind all of this, is it not the same logic that appears to be at work: that of surviving, using its mechanisms for repairs? This is incidentally a temporal logic, it would appear, with a hierarchy of clocks, as found in the latest generation chips.

Response by Philippe Kourilsky

I was indeed very careful with this notion of quanta, and to be honest I am not very comfortable with it. The issue for me is to know how we could imagine the nodes of biological networks. These networks simultaneously have a "wave-like" aspect, since there are interactions that are characterized virtually by overall physicochemical constants (affinity, chemical potential, etc.) and a "corpuscular" side with manufactured proteins which, for their part, are entities that could not be any more concrete. It is in thinking of this, with this idea of a double function, that we could try to make a parallel with the notions of waves/particles from physics. But that remains to be explored, by physicists and maybe programmers; for the moment, then, it does not extend very far.

My conviction, today, is that a cell is a machine that can be found in a very large number of possible states, which we still do not know how to investigate and/or experiment with properly. As I have said before, a cell is somewhat like a satellite with antennas that, depending on the signals they capture, will activate to implement one or another task. As I see it, even if it is not stimulated, it will oscillate between several states, currently still not counted or characterized, in a discrete or almost continuous fashion. With what modes of fluctuation, we do not yet know, but we observe oscillations. If we analyze individual cells meant to be completely identical, we find a surprising diversity of states. Is it simply background noise? The real issue is knowing if this noise gets used rather than remaining useless.

My hypothesis, relative to the immune system, is that it is an important property of circulating cells, while combined cells control each other, and thus the number of their possible states diminishes greatly. They can provide an almost stable reference to the circulating cells which come to redefine themselves as a result of random encounters. The immune system works thanks to a large number of circulating cells that irrigate the entire organism. I proposed that these fluctuations are used by the immune system to expand its capacity to recognize. That is what we can say about this somewhat prospective issue, but we must keep in mind that the immune system works by constantly multiplying and/or destroying physical elements in a controlled manner.

REMARK BY JACQUES PRINTZ.– *In the latest generation SOCs (Systems On Chip) currently containing 4–5 billion transistors, and something in the range of 50 billion in the near future, there is an entire hierarchy of clocks to form the rhythm of the cadence of the chip and put the temporarily inactive parts of the chip on sleep mode in order to save energy and prevent it from overheating*[9]*. The right image would be a hierarchy of loops interlinked with each other and resynchronized periodically so*

9 In 2021, Apple engineers developed an SOC of 432 mm^2 containing 55 billion components with engravings of 2–3 nanometers.

that the circuit does not go haywire, continuing through to the highest level of the computer stack, with specific characteristics at each of the levels. This is a major issue for asynchronous distributed architectures where we progressively shift from organic clocks, cadenced by the physical processes in the silicon, to semantic clocks, cadenced by what users do in the software.

Question no. 6: complexity

To finish this interview, we cannot forget to mention complexity in reference to the title of your book, which appears to echo the (already old) work by Jacques Monod, *Chance and Necessity*[10]. Making a link between chance and complexity is very interesting, even if that must have surprised your biologist colleagues; and especially, it seems to me, if we include what you qualify as saturable/non-saturable. In systems engineering, there is an engineering of "chance" that is used for survival by ensuring proper functioning, but perhaps it would be better to speak of randomness. Error correction codes, invented by Shannon, can correct "noise" from the environment because the "noise" is random. The chance of physicists like Boltzmann becomes a beneficial property when we know how to integrate it; this is perfectly clear when we study the anatomy of the computer stack, which is the most striking example of it, and that moreover everyone has to hand. It is but one example, but we can imagine a general theory that would give shape to all of this using an ad hoc language. It seems obvious that the living world has developed analogous strategies. A living organism with its hundreds of billions of interacting cells cannot be determined by "chance" as Monod understood it; for it already no longer works for the artifacts made by engineers since that time. The smallest of the latest generation microprocessors and their memories contain billions of components[11]. We are still way off the mark if we compare it with the living, but we are, however, certain that if we were to leave it up to chance, the average time of proper functioning for these systems would only be a few minutes! Here again, it is von Neumann that gave us the key in his monograph *Probabilistic Logic and the Synthesis of Reliable Organisms From Unreliable Components* (Von Neumann 1961) – it is both great and prophetic, since it is organization that compensates for "noise" – but I have almost never seen his work cited among the community of biologists, or at least those whom I know, which is, very few. Can you enlighten us on this major point to conclude, and give us your vision of complexity via the immune system, a true "miracle" of engineering for an engineer like me?

---

10 Published by Alfred A. Knopf in 1971.

11 21 billion for the latest generation chip of the NVIDIA QUADRO GV100 computer, with an engraving technology of 12 nanometers.

Response by Philippe Kourilsky

It is a vast subject and we have already dealt with certain aspects of it earlier on. In my opinion, evolution always creates more complexity. Robustness is necessarily a driver for selection within a species, since, if an individual is more robust, their chances of survival increase. In mankind, this is more complicated because robustness can also come from the brain. From this point of view, humankind has demonstrated that they are more robust than monkeys, which have not invented medicine or vaccines.

Over the long term, in the evolution of species, we observe this evolution towards more complexity within genomes. In human genomes themselves, we see traces of recent evolutions that sometimes only go back to a few thousand years. We can correlate certain genetic modifications with changes in diet and social organization, for example, in the transition from hunter-gatherers to farmer-breeders. These correlations do not constitute proof, but they are suggestive.

For me, it would be worthwhile focusing our considerations on complexity around the notion of robustness. In the history of living things, we can see that there are additions that develop randomly, and which, if they provide a competitive advantage in terms of survival, will endure and be transmitted to future generations. It seems to me that the more complex a system is, the more chances it will have to make errors, and the more robustness will become a major parameter in its evolution. There is a sort of back and forth between function and robustness. Infectious agents, in particular, intervene in this spiraling motion. This leads to a joint evolution of genes in the organism defending itself and in the infectious agents attacking it. Because of this, the system always moves towards an increase in complexity while maintaining a balance between biological cost and quality control or repair mechanisms.

I would again like to emphasize the fact that to build reliable automata with non-reliable elements, to use von Neumann's formula which you mentioned, we must consider two things: the first is that living systems have mechanisms for identifying and recycling waste, with its physical elimination, and the second is the compromises that the organism makes given what have been called its "specificities". The world of the living is a world of intersecting reactions. Reliability depends on the threshold of tolerance relative to these intersecting reactions. Biological systems do not aspire to perfection. This is not possible, neither theoretically nor practically. The case of DNA replication shows that a rate of imperfections as low as 10^{-9} (especially when they accumulate) is not without consequences, in either the life of organisms or in their evolution.

To grasp this kind of consideration regarding reliability and the way randomness comes into all of this, that is, groups of hundreds of billions of cells, something in the order of $10^{13}/10^{14}$ as we saw, I think that as biologists, we have a real mathematical problem. Mathematics is understood here as a language, which I do not really know how to formulate, but my allusion to the biology of quanta was also an allusion to quantum mechanics and thus related mathematics, of which, incidentally, von Neumann was the principal artisan, with Hilbert.

Returning to genetic engineering, which not so long ago was called "genetic manipulations", the frequently expressed fear of arriving at a genetic engineering of humankind does not seem entirely unfounded to me over the long term! Let us make a parallel with this other science: chemistry. Its evolution has been remarkable in the sense that chemists today work essentially on objects they have created or are creating entirely. From this we can imagine a future where biology would above all concern itself with biological beings that were themselves also entirely created. This comparison gives us food for thought. The fantasies about "mad scientists" that create/fabricate something living have a future since what we call synthetic biology today is headed in this direction. Is it on the basis of synthetic biology that we can imagine systems that will integrate classical engineering of the "biological computer" type, which we hear about here and there? We will see. All of this raises worrying ethical issues.

This also reminds me of the logic of self-teaching in *probably approximately correct*, as described in the book by Leslie Valiant, the PAC logic, which you tell me you appreciated greatly[12]. Basically, there are algorithms that are given an objective, capable of approximately measuring the distance from the objective, with a memory of what has been done, to find a pathway from the present state to the one corresponding to the future objective being sought out. Through this, we can develop small, ultra-efficient and low-cost algorithms. This is a possibly essential logic in biology that correctly describes many biological processes. For example, for the configuration of proteins, when we see the mess of atoms that must come together to constitute a functional protein in the end, we might think that the strategy used is rather efficient after all, even if it creates a fair amount of waste. Which brings us to two essential issues in biology, first the question of finality (teleology), then that of energy and the minimum amounts needed for the system to remain, economically speaking, efficient so it can survive hazards. A sort of constant?

REMARK BY JACQUES PRINTZ.– *When you speak to us about a "mess of atoms", I cannot help but think of this image (shown on the right-hand side of Figure 6.6) taken from the book by Robert Laughlin,* A Different Universe, *which stands out to*

12 The complete title of this excellent book is *Probably Approximately Correct – Nature's Algorithms for Learning and Prospering in a Complex World*; Leslie Valiant is a first-rate British programmer whose work won him the Turing Award in 2010.

any engineer with basic training[13]*. To make a chicken from a mess of atoms, at the very least, an entire set of construction norms is needed, including validation protocols for every level, especially the modules, in order to go from the initial egg, a unique cell that is a fusion of an ovum and a spermatozoid, to an organism that will include hundreds of billions of them!*

I would like to evoke one last essential notion for the science of the living and specifically for immunology. It is the issue of distinguishing/discriminating between the self and the non-self. This comes down to asking ourselves what the self is, then what the non-self is, and moreover what the non-self is with respect to the self.

This led me to propose that the self is none other than conformity to a set of "norms" or rules that must be materialized in the organism and in all the elements that take part in the regulation and controls that we have spoken of non-stop since the beginning. This set of norms is obviously an element of the overall complexity. We can thus define the non-self by its inverse, that is, everything that does not correspond to these norms.

What are these norms? They are rules of a biochemical and functional nature (e.g. proteins are molecules that must respect the rules of chemistry). The non-self is therefore whatever is not structurally and/or functionally normal. That brings us to the definition of physiology given by Claude Bernard, which might appear circular, but that we must recall is foundational in medicine. The crux for biologists is thus to establish the catalogue of norms. In other words, in a living organism, robustness is the verification of the catalogue of norms.

In this sense, we can say that biologists are coherent with what you have explained in your book on systemics, with the functional aspects and the non-functional aspects, if I have understood correctly. The problem of transposing notions that come from the sphere of artificial systems engineering to that of biological systems boils down to a real problem of language, in the strongest sense of the word. We must use words that have been carefully defined, in a sufficiently abstract way but not too much, so we can bridge two very different realities. It is far from being easy, hence my idea for new mathematics. When there are ideas that appear to be interesting, we will have to experiment anyway in order to decide, and, even when that is possible, it can take a lot of time.

13 Robert Laughlin is a physicist specializing in condensed matter (solid state physics) and whose research won him the Nobel Prize in 1998. Along with Philip Warren Anderson, he supports the rather revolutionary thesis of emergence, of which the semi-conductors used in computers are an excellent illustration; see Part 1.

Simulation techniques are beginning to fascinate biologists. For example, studies concerning the "virtual patient" integrate annotated functional blueprints in complex representations (like the ones I spoke of to you at the very start) using what we already know about the way the organism works. The annotations could be specific to a particular patient. These models will help surgeons train, but they will also serve in testing new medications. There is perhaps, in this, a new field to explore where bridges can be found between our disciplines. It provides us, temporarily, with a conclusion…

Convergence of disciplines? Under what condition(s)?

In developing this conclusion, I am under the impression that what brings us together the most is perhaps this issue of errors with the PAC logic, a logic that implies modularity, which is the fundamental quantum of information sciences; perhaps this is what we could conclude with?

As far as I am concerned, that is what our dialogue brings to mind.

What you evoke with respect to the relationship of robustness versus complexity, as well as the essential capacity to control, all of this built upon an understanding of the randomness that living systems must face, reminds me of what happened at the start of the 1980s. At this key turning point, computers were still subject to random errors due to their environment, essentially errors in memory parity, with an average frequency of one error for a few hours of work. The countermeasure consisted of having "photos" of the state of the memory taken periodically by the operating system, which meant that in the case of an error that had to be diagnosed, it was enough to restart the computer from the last photo, called a snapshot in the jargon. The error was simply made invisible.

The 1980s were the years when microelectronics and the integration of components were beginning; transistors became an abundant resource, which meant that we were able to go from simple control of parity to error correction codes, invented by Claude Shannon 30 years earlier. The same thing happened in the development of communication means, with data transport networks as well as the implementation of parallel processing. A good architecture of the abundance of resources paired with an understanding of the random phenomena in each of the different levels of organization progressively led to the current computers whose errors, still just as numerous if not more given the increase in complexity, are completely hidden from users.

The magic of randomness is its definition itself. A random error will never repeat identically; otherwise, it would not be random. We are, in the end, protected by the immensity of the underlying combinatorials. Engineering the robustness of artificial systems intelligently combines chance and complexity, which brings us back to the title of your book!

Response by Philippe Kourilsky

I very much appreciate what you are saying, which clearly shows the possibility of convergences between our two disciplines, something that we had both suspected in starting this dialogue. Throughout our discussion, I have tried to show the parallels and similarities all the while identifying several significant differences. The depth of these is not a deal-breaker. It only increases the utility of exploring convergences further. Our exercise both thrilled and frustrated me. Thrilled, because I am coming away from this, like you, reassured by the idea that there is an important and new field to work on. Is it not at the interfaces between disciplines that the most innovative sparks are often produced? Frustrated, because the breadth of the task is considerable. I have the feeling that it surpasses me, and this is for several reasons. The first is that our disciplines are already, in fact, groups of disciplines that are difficult to learn in and of themselves. The second is that each one of them has undergone its own evolution and developed its own language. We belong to different tribes, which have mutual sympathy, and which have the intuition of being able to hybridize. But we must still prove that; as with a horse and a donkey, we have not diverged to the point of losing fertility. Now, mutual understanding is not easy, not only because concepts and knowledge are, in part, different, but also because the dialects are structured differently. The words do not always have the same meaning for you and for me, and the first thing to do would be to build a dictionary that would allow the tribes to communicate. In our dialogue we have sketched it out, while at the same time remarking that this great and important work should be led by someone who has a profound understanding of both languages and of both cultures. Research is being done at these interfaces, and university training is emerging. I have no doubt that our modest exchange carries the seed of exciting developments that will soon be taking shape.

List of Acronyms

AC/TC	Algorithmic Complexity/Textual Complexity
AF	Architecture Framework; found in many acronyms, such as TOGAF (The Open Group Architecture Framework)
AU	Active Unity; the indivisible core of a project team, composed of five to ten people
BPML/BPMN	Business Process Modeling Language/Notation
C2/C3/C4ISR C4ISTAR	Acronyms used to represent the group of military functions designated by C4 (Command, Control, Communications, Computers), I (Military Intelligence) and STAR (Surveillance, Target Acquisition and Reconnaissance) to allow the coordination of operations
CCCAM	*Cybernetics or Control and Communication in the Animal and the Machine* (seminal work by Wiener)
CCITT	International Telegraph and Telephone Consultative Committee; until 1993. It dealt with technical questions and normalization and was replaced by the ITU
CERN	The European Organization for Nuclear Research (*Centre Européen de Recherche Nucléaire*)

NBIC/STIC/ICT	Nanotechnology, Biology, Informatics, Cognition/Science and Technology of Information and Communication/Information and Communications Technology
NCS	Network Centric Systems; systems whose architectural heart is the network
NTDS	Navy Tactical Data System; the first naval system in the US Navy
PAC	*Probably Approximately Correct* (title of the book by Valiant)
PESTEL	Political, Economic, Social, Technological, Ecological, Legal
PSTN	Public Switched Telephone Network
QI	Quantity of Information; the formula used by Shannon
RAS	Reliability, Availability, Serviceability
SADT/SART	Structured Analysis and Design Techniques (Real Time); the basis of the IDEF language
SAGE	Semi-Automatic Ground Environment; the ancestor of all air control systems
SCADA	Supervisory Control And Data Acquisition; any C4ISTAR system has a SCADA peripheral
SCCOA	*Système de Commandement et de Conduite des Opérations Aériennes*; the system for managing air combat operations
SENIT	*Système d'Exploitation Navale des Informations Tactiques*; the system of tactical information for the navy, based on the American NTDS
SI/SIC/SIO	*Système d'Information/Système d'Information et de Communication/Système d'Information Opérationnel* (*dans le domaine de la défense et de la sécurité*); (operational) information and communication systems used in defense and security
SLA	Service Level Agreement; or service contract
SoS	System of Systems

STRIDA	*Système de Traitement et de Représentation des Informations de Défense Aérienne*; based on the SAGE system for processing and representing air defense data
SysML	Systems Modeling Language; extension based on UML
TCO	Total Cost of Ownership; in project management
TOGAF	The Open Group Architecture Framework
TPS	Transactions Per Second; measures the speed of a system in number of interactions per second
UML	Unified Modeling Language; promoted by the OMG (Object Management Group)
USE	User, System, Engineering; this is Simondon's triplet
VHDL	Very High Description Language; for the description of VLSI, inspired by Ada, and integrating parallelism
VHV	Very High Voltage cables in the electrical network; between 400 and 700,000 volts
VLSI/LSI	Very Large-Scale Integration; from the 1990s, when the number of transistors per chip rose to dozens of millions
XML/SGML	eXtended Markup Language; the successor of the Standard Generalized Markup Language, which appeared with the Internet and the Web

References

Many works have been cited and/or used during the writing of this book. Their usefulness is essentially of a historical nature for those who wish to deepen and understand the genesis of the concepts presented in this work and the previous one. You do not have to read everything. This is why we have selected a short list of around twenty books that can be considered essential in the library of a systems engineer.

Anderson, P.W. (1972). More is different. *Science*, 177(4047).

Brillouin, L. (1959a). *Science and Information Theory*. Dover Publications, New York.

Brillouin, L. (1959b). *Vie, matière et observation*. Albin Michel, Paris.

Caseau, Y. (2005). *Urbanisation et BPM*. Dunod, Paris.

Cohen-Tannoudji, G. and Spiro, M. (2013). *Le boson et le chapeau mexicain*. Gallimard, Paris.

Copeland, J. (2004). *The Essential Turing*. Clarendon Press, Oxford.

Dijkstra, E. (1968). Cooperating sequential processes. In *Programming Languages: NATO Advanced Study Institute*, Genuys, F. (ed.). Academic Press, London.

Forrester, J. (1961). *Industrial Dynamics*. MIT Press, Cambridge.

Forrester, J. (1971). *Principles of Systems*. Pegasus, Cambridge.

Hamming, R. (1980). *Coding and Information Theory*. Prentice Hall, Hoboken.

Humphrey, W. (1989). *Manage the Software Process*. Addison-Wesley, Boston.

IEEE (2019). *System Engineering Standards Collection*. Standard 24748.

INCOSE (2015). *Systems Engineering Handbook: A Guide for System Life Cycle Processes and Activities*, 4th edition. Wiley, New York.

ISO/IEC (2015). *Systems Engineering Standard Covering Processes and Lifecycle Stages.* Standard 15288.

Korzybski, A. (1994). *Science and Sanity*, 5th edition. Institute of General Semantic, New York.

Kourilsky, P. (2014). *Le jeu du hasard et de la complexité.* Odile Jacob, Paris.

Krob, D. (2017). CESAMES Systems Architecting Method. Working paper, CESAMES, Paris.

Lesourne, J. (1977). *Les systèmes du destin.* Dalloz, Paris.

Martin, J. (1989). *Information Engineering.* Prentice Hall, Hoboken.

NASA (2017). System Engineering Handbook. Working document, NASA, Washington.

von Neumann, J. (1961). *Collected Works.* Pergamon Press, Oxford.

Printz, J. (2006). *Écosystème des projets informatiques.* Hermes-Lavoisier, Paris.

Printz, J. (2013a). *Architecture logicielle.* Dunod, Paris.

Printz, J. (2013b). *Estimation des projets de l'entreprise numérique.* Hermès-Lavoisier, Paris.

Printz, J. (2018). *Survivrons-nous à la technologie.* Les Acteurs du savoir, Paris.

Printz, J. (2020). *System Architecture and Complexity: Contribution of Systems of Systems to Systems Thinking.* ISTE Ltd, London and John Wiley & Sons, New York.

Rechtin, E. (2000). *The Art of Systems Architecting.* CRC Press, Boca Raton.

Senge, P. (1990). *The Fifth Discipline.* Random House, New York.

Shannon, C. and Weaver, W. (1959). *The Mathematical Theory of Communication.* University of Illinois Press, Champaign.

Sifakis, J. (2013). Rigorous system design. *Foundations and Trends in Electronic Design Automation*, 6(4), 293–362.

Simon, H. (1996). *The Sciences of the Artificial.* MIT Press, Cambridge.

Simondon, G. (1958). *Du mode d'existence des objets techniques.* Aubier, Paris.

Valiant, L. (2013). *Probably Approximately Correct: Nature's Algorithms for Learning and Prospering in a Complex World.* Basic Books, New York.

Vaughan, D. (1997). *The Challenger Launch Decision.* University of Chicago Press, Chicago.

Weinberg, G. (1975). *An Introduction to General Systems Thinking.* Dorset House, New York.

Wiener, N. (1948). *Cybernetics or Control and Communication in the Animal and the Machine.* MIT Press, Cambridge.

Index

A, C

D, E

F, H, I

L, M

2020

BRON Jean-Yves
System Requirements Engineering

KRYSINSKI TOMASZ, MALBURET FRANÇOIS
Energy and Motorization in the Automotive and Aeronautics Industries

PRINTZ Jacques
System Architecture and Complexity: Contribution of Systems of Systems to Systems Thinking

2019

ANDRÉ Jean-Claude
Industry 4.0: Paradoxes and Conflicts

BENSALAH Mounir, ELOUADI Abdelmajid, MHARZI Hassan
Railway Information Modeling RIM: The Track to Rail Modernization

BLUA Philippe, YALAOU Farouk, AMODEO Lionel, DE BLOCK Michaël, LAPLANCHE David
Hospital Logistics and e-Management: Digital Transition and Revolution

BRIFFAUT Jean-Pierre
From Complexity in the Natural Sciences to Complexity in Operations Management Systems
(Systems of Systems Complexity Set – Volume 1)

BUDINGER Marc, HAZYUK Ion, COÏC Clément
Multi-Physics Modeling of Technological Systems

FLAUS Jean-Marie
Cybersecurity of Industrial Systems

JAULIN Luc
Mobile Robotics – Second Edition Revised and Updated

KUMAR Kaushik, DAVIM Paulo J.
Optimization for Engineering Problems

TRIGEASSOU Jean-Claude, MAAMRI Nezha
Analysis, Modeling and Stability of Fractional Order Differential Systems 1: The Infinite State Approach
Analysis, Modeling and Stability of Fractional Order Differential Systems 2: The Infinite State Approach

VANDERHAEGEN Frédéric, MAAOUI Choubeila, SALLAK Mohamed, BERDJAG Denis
Automation Challenges of Socio-technical Systems

2018

BERRAH Lamia, CLIVILLÉ Vincent, FOULLOY Laurent
Industrial Objectives and Industrial Performance: Concepts and Fuzzy Handling

GONZALEZ-FELIU Jesus
Sustainable Urban Logistics: Planning and Evaluation

GROUS Ammar
Applied Mechanical Design

LEROY Alain
Production Availability and Reliability: Use in the Oil and Gas Industry

MARÉ Jean-Charles
Aerospace Actuators 3: European Commercial Aircraft and Tiltrotor Aircraft

MAXA Jean-Aimé, BEN MAHMOUD Mohamed Slim, LARRIEU Nicolas
Model-driven Development for Embedded Software: Application to Communications for Drone Swarm

MBIHI Jean
Analog Automation and Digital Feedback Control Techniques
Advanced Techniques and Technology of Computer-Aided Feedback Control

MORANA Joëlle
Logistics

SIMON Christophe, WEBER Philippe, SALLAK Mohamed
Data Uncertainty and Important Measures
(Systems Dependability Assessment Set – Volume 3)

TANIGUCHI Eiichi, THOMPSON Russell G.
City Logistics 1: New Opportunities and Challenges
City Logistics 2: Modeling and Planning Initiatives
City Logistics 3: Towards Sustainable and Liveable Cities

ZELM Martin, JAEKEL Frank-Walter, DOUMEINGTS Guy, WOLLSCHLAEGER Martin
Enterprise Interoperability: Smart Services and Business Impact of Enterprise Interoperability

2017

ANDRÉ Jean-Claude
From Additive Manufacturing to 3D/4D Printing 1: From Concepts to Achievements
From Additive Manufacturing to 3D/4D Printing 2: Current Techniques, Improvements and their Limitations
From Additive Manufacturing to 3D/4D Printing 3: Breakthrough Innovations: Programmable Material, 4D Printing and Bio-printing

ARCHIMÈDE Bernard, VALLESPIR Bruno
Enterprise Interoperability: INTEROP-PGSO Vision

CAMMAN Christelle, FIORE Claude, LIVOLSI Laurent, QUERRO Pascal
Supply Chain Management and Business Performance: The VASC Model

FEYEL Philippe
Robust Control, Optimization with Metaheuristics

MARÉ Jean-Charles
Aerospace Actuators 2: Signal-by-Wire and Power-by-Wire

POPESCU Dumitru, AMIRA Gharbi, STEFANOIU Dan, BORNE Pierre
Process Control Design for Industrial Applications

RÉVEILLAC Jean-Michel
Modeling and Simulation of Logistics Flows 1: Theory and Fundamentals
Modeling and Simulation of Logistics Flows 2: Dashboards, Traffic Planning and Management
Modeling and Simulation of Logistics Flows 3: Discrete and Continuous Flows in 2D/3D

2016

ANDRÉ Michel, SAMARAS Zissis
Energy and Environment
(Research for Innovative Transports Set – Volume 1)

AUBRY Jean-François, BRINZEI Nicolae, MAZOUNI Mohammed-Habib
Systems Dependability Assessment: Benefits of Petri Net Models
(Systems Dependability Assessment Set – Volume 1)

BLANQUART Corinne, CLAUSEN Uwe, JACOB Bernard
Towards Innovative Freight and Logistics
(Research for Innovative Transports Set – Volume 2)

COHEN Simon, YANNIS George
Traffic Management
(Research for Innovative Transports Set – Volume 3)

MARÉ Jean-Charles
Aerospace Actuators 1: Needs, Reliability and Hydraulic Power Solutions

REZG Nidhal, HAJEJ Zied, BOSCHIAN-CAMPANER Valerio
Production and Maintenance Optimization Problems: Logistic Constraints and Leasing Warranty Services

TORRENTI Jean-Michel, LA TORRE Francesca
Materials and Infrastructures 1
(Research for Innovative Transports Set – Volume 5A)
Materials and Infrastructures 2
(Research for Innovative Transports Set – Volume 5B)

WEBER Philippe, SIMON Christophe
Benefits of Bayesian Network Models
(Systems Dependability Assessment Set – Volume 2)

YANNIS George, COHEN Simon
Traffic Safety
(Research for Innovative Transports Set – Volume 4)

2015

AUBRY Jean-François, BRINZEI Nicolae
Systems Dependability Assessment: Modeling with Graphs and Finite State Automata

BOULANGER Jean-Louis
CENELEC 50128 and IEC 62279 Standards

BRIFFAUT Jean-Pierre
E-Enabled Operations Management

MISSIKOFF Michele, CANDUCCI Massimo, MAIDEN Neil
Enterprise Innovation

2014

CHETTO Maryline
Real-time Systems Scheduling
Volume 1 – Fundamentals
Volume 2 – Focuses

DAVIM J. Paulo
Machinability of Advanced Materials

ESTAMPE Dominique
Supply Chain Performance and Evaluation Models

FAVRE Bernard
Introduction to Sustainable Transports

GAUTHIER Michaël, ANDREFF Nicolas, DOMBRE Etienne
Intracorporeal Robotics: From Milliscale to Nanoscale

MICOUIN Patrice
Model Based Systems Engineering: Fundamentals and Methods

MILLOT Patrick
Designing Human–Machine Cooperation Systems

NI Zhenjiang, PACORET Céline, BENOSMAN Ryad, RÉGNIER Stéphane
Haptic Feedback Teleoperation of Optical Tweezers

OUSTALOUP Alain
Diversity and Non-integer Differentiation for System Dynamics

REZG Nidhal, DELLAGI Sofien, KHATAD Abdelhakim
Joint Optimization of Maintenance and Production Policies

STEFANOIU Dan, BORNE Pierre, POPESCU Dumitru, FILIP Florin Gh., EL KAMEL Abdelkader
Optimization in Engineering Sciences: Metaheuristics, Stochastic Methods and Decision Support

2013

ALAZARD Daniel
Reverse Engineering in Control Design

ARIOUI Hichem, NEHAOUA Lamri
Driving Simulation

CHADLI Mohammed, COPPIER Hervé
Command-control for Real-time Systems

DAAFOUZ Jamal, TARBOURIECH Sophie, SIGALOTTI Mario
Hybrid Systems with Constraints

FEYEL Philippe
Loop-shaping Robust Control

FLAUS Jean-Marie
Risk Analysis: Socio-technical and Industrial Systems

FRIBOURG Laurent, SOULAT Romain
Control of Switching Systems by Invariance Analysis: Application to Power Electronics

DAVIM J. Paulo
Mechatronics

DAVIM J. Paulo
Wood Machining

GROUS Ammar
Applied Metrology for Manufacturing Engineering

KOLSKI Christophe
Human–Computer Interactions in Transport

LUZEAUX Dominique, RUAULT Jean-René, WIPPLER Jean-Luc
Complex Systems and Systems of Systems Engineering

ZELM Martin, *et al.*
Enterprise Interoperability: IWEI2011 Proceedings

2010

BOTTA-GENOULAZ Valérie, CAMPAGNE Jean-Pierre, LLERENA Daniel, PELLEGRIN Claude
Supply Chain Performance: Collaboration, Alignment and Coordination

BOURLÈS Henri, GODFREY K.C. Kwan
Linear Systems

BOURRIÈRES Jean-Paul
Proceedings of CEISIE'09

CHAILLET Nicolas, REGNIER Stéphane
Microrobotics for Micromanipulation

DAVIM J. Paulo
Sustainable Manufacturing

GIORDANO Max, MATHIEU Luc, VILLENEUVE François
Product Life-Cycle Management: Geometric Variations

LOZANO Rogelio
Unmanned Aerial Vehicles: Embedded Control

LUZEAUX Dominique, RUAULT Jean-René
Systems of Systems

VILLENEUVE François, MATHIEU Luc
Geometric Tolerancing of Products

2009

DIAZ Michel
Petri Nets: Fundamental Models, Verification and Applications

OZEL Tugrul, DAVIM J. Paulo
Intelligent Machining

PITRAT Jacques
Artificial Beings

2008

ARTIGUES Christian, DEMASSEY Sophie, NÉRON Emmanuel
Resources–Constrained Project Scheduling

BILLAUT Jean-Charles, MOUKRIM Aziz, SANLAVILLE Eric
Flexibility and Robustness in Scheduling

DOCHAIN Denis
Bioprocess Control

LOPEZ Pierre, ROUBELLAT François
Production Scheduling

THIERRY Caroline, THOMAS André, BEL Gérard
Supply Chain Simulation and Management

2007

DE LARMINAT Philippe
Analysis and Control of Linear Systems

DOMBRE Etienne, KHALIL Wisama
Robot Manipulators

LAMNABHI Françoise *et al.*
Taming Heterogeneity and Complexity of Embedded Control

LIMNIOS Nikolaos
Fault Trees

2006

FRENCH COLLEGE OF METROLOGY
Metrology in Industry

NAJIM Kaddour
Control of Continuous Linear Systems

Printed and bound by CPI Group (UK) Ltd, Croydon, CR0 4YY

16/09/2026

14963083-0006